Palgrave Studies in Digital Business & Enabling Technologies

Series Editors
Theo Lynn
Irish Institute of Digital Business
DCU Business School
Dublin, Ireland

John G. Mooney
Graziadio Business School
Pepperdine University
Malibu, CA, USA

This multi-disciplinary series will provide a comprehensive and coherent account of cloud computing, social media, mobile, big data, and other enabling technologies that are transforming how society operates and how people interact with each other. Each publication in the series will focus on a discrete but critical topic within business and computer science, covering existing research alongside cutting edge ideas. Volumes will be written by field experts on topics such as cloud migration, measuring the business value of the cloud, trust and data protection, fintech, and the Internet of Things. Each book has global reach and is relevant to faculty, researchers and students in digital business and computer science with an interest in the decisions and enabling technologies shaping society.

More information about this series at
http://www.palgrave.com/gp/series/16004

Theo Lynn
John G. Mooney
Pierangelo Rosati • Grace Fox
Editors

Measuring the Business Value of Cloud Computing

palgrave
macmillan

Editors
Theo Lynn
Irish Institute of Digital Business
DCU Business School
Dublin, Ireland

John G. Mooney
Graziadio Business School
Pepperdine University
Malibu, CA, USA

Pierangelo Rosati
Irish Institute of Digital Business
DCU Business School
Dublin, Ireland

Grace Fox
Irish Institute of Digital Business
DCU Business School
Dublin, Ireland

ISSN 2662-1282   ISSN 2662-1290 (electronic)
Palgrave Studies in Digital Business & Enabling Technologies
ISBN 978-3-030-43197-6   ISBN 978-3-030-43198-3 (eBook)
https://doi.org/10.1007/978-3-030-43198-3

This Palgrave Macmillan imprint is published by the registered company Springer Nature Switzerland AG.
The registered company address is: Gewerbestrasse 11, 6330 Cham, Switzerland

PREFACE

The fourth volume in the Palgrave Studies in *Digital Business & Enabling Technologies* aims to advance knowledge and offer multidisciplinary insight into the area of business value associated with enabling technologies. Specifically, the book seeks to better understand approaches for conceptualising and measuring business value from the implementation of cloud computing technologies. The importance of demonstrating the value achieved from IT investments is long established in the Computer Science (CS) and Information Systems (IS) literature. However, the complexity and convergence of next generation technologies, including cloud computing, presents new challenges and opportunities for demonstrating how IT investments lead to business value. Recent reviews of extant literature highlight the need for multi-disciplinary research which both explores and further develops the conceptualization of value in cloud computing research, and research which investigates how IT value manifests itself across the chain of service provision and in inter-organizational scenarios.

At the heart of business value research is the desire to understand how information technology can improve the performance of an organisation. Due to the multi-disciplinary nature of the business value domain, the extant literature is characterised by a broad range of methodologies including qualitative case studies and quantitative calculations of value, as well as a myriad of different IT artefacts across varying units of analysis from a business process, unit, organisational, inter-organisational, and value chain levels. Traditionally, business value research was concerned with providing a justification for IT investments. Recent advances in information technologies and the advent of so-called third platform technologies (e.g.

mobile, social, Big Data analytics, Internet of Things, and as cloud computing technologies) enable shifts in the distribution of costs over time based on resource allocation as opposed to the large upfront investment required in traditional system implementations such as enterprise resource planning systems (ERP). Furthermore, the flexible and interdependent nature of cloud computing may introduce new intangible benefits. It is thus important to examine the different approaches to measuring the value of cloud computing investments across the various cloud service provision models and deployment models.

In response to the call for multi-disciplinary research, contributors to the book have been drawn from an international group of scholars in IS, CS, and accounting. Measuring the Business Value of Cloud Computing reviews the state of the art from these varying perspectives to detail the prevailing techniques for measuring business value for cloud computing across a variety of scenarios and illustrative mini-cases. Chapter 1 begins by laying the foundational justification for measuring business value in the cloud computing context by highlighting the growth in cloud computing expenditure. The introductory chapter reviews the established measures of business value and seeks to determine the relevance of these measures to the cloud computing context. The traditional measurement of IT business value involves the calculation of different metrics including Net Present Value (NPV), Return on Investment (ROI), Payback Period, Internal Rate of Return (IRR), Economic Value Added (EVA), and Total Cost of Ownership (TCO). However, cloud computing introduces new metrics such as resilience, speed of deployment, scalability, and organisational agility, as well as new intangible benefits which make measurement of value more difficult. To overcome this, the authors suggest the potential of assessment approaches such as scoring, value linking, and value acceleration or holistic approaches such as the Business Value Index (BVI) grid. The authors conclude by highlighting the importance of measuring not only the value but the realisation of proposed benefits following the adoption of cloud computing.

Building on the broad foundation laid by Chap. 1, the subsequent two chapters focus on specific cloud provision models. When an organisation is considering the adoption of cloud computing, it is imperative to determine what cloud service model meets the organisation's needs. Chapter 2 focuses on Infrastructure-as-a-Service (IaaS) and Platform-as-a-Service (PaaS) and discusses the suitability of calculating Return on Investment (ROI) as a measurement of value as opposed to the most used measure of

TCO. The chapter details a six step process for calculating ROI which encompasses both costs and benefits. The proposed process is illustrated by calculating ROI in a case study of an IaaS migration project by a global financial services organisation. Moving on from IaaS and PaaS, Chap. 3 focuses on Software-as-a-Service (SaaS), the dominant cloud service provision model. Taking a broader perspective, the chapter focuses on identifying the business model payoffs fostered by SaaS technologies. Adopting a case study approach to compare two large, multi-national incumbent IT service providers leading SaaS provision, the chapter identifies six tangible payoffs categorised as economic, business and transformative payoffs.

Moving on from the perspectives of customers and providers detailed in Chaps. 2 and 3 respectively, it is important to understand the wider cloud computing landscape and the stakeholders that operated or affected by it. The subsequent three chapters adopt broader approaches to exploring the role of value in cloud computing across multiple organisations. Chapter 4 deals with another important player in the cloud landscape namely B2B cloud marketplaces. The chapter focuses on the role of B2B cloud marketplaces within the cloud service brokerage (CSB) landscape. The chapter discusses B2B cloud marketplaces both in terms of the structural level and the functional level detailing the characteristics and benefits of B2B cloud marketplaces such as ease-of-use, ease-of-integration, enhanced security, increased manageability, faster implementation, and cost reduction. Leveraging two mini case studies to represent the two types of B2B cloud marketplaces (business application marketplace and the API marketplace), the chapter details how cloud customers can utilise both marketplaces to derive measurable value.

When considering adopting cloud computing, cloud consumers must first identify their requirements. Chapter 5 details the ten prevailing cloud deployment models including public clouds, private clouds, and federated clouds. Each cloud deployment model is characterised by differing costs and benefits. To aid cloud consumers in differentiating between different cloud deployment models and guide the identification of the appropriate deployment model, Chap. 5 develops and presents a comprehensive cost model which details the pertinent cost factors at play and the underlying economic models. Building upon this discussion, the chapter identifies the potential of federated clouds for overcoming some of the economic challenges and develops a ten-step use case scenario for applying an economic model in cloud federation deployments implemented through three modules of service placement, accounting, and revenue sharing. To round off the exploration of value in terms of cloud computing, Chap. 6 focuses on the wider digital ecosystem and the role of digital platforms in influencing the value creation structure of ecosystems. Specifically, the chapter reviews the role of power asymmetry in impacting value creation on a digital platform. Using the example of a cloud-based gaming platform, the chapter details the direct and indirect value that network actors create for each other and the end customer.

The final chapter builds upon a recent literature review of the extant knowledge base on measuring the impact of cloud computing investments. While the previous chapters seek to address gaps in our knowledge identified in this review around approaches to measurement, this chapter focuses on emerging paradigms which may impact cloud computing including the heterogeneity, fog and edge computing, and machine learning and artificial intelligence for IT operations (AIOps). The chapter explores how these technological advancements may further complicate the measurement of business value derived from implementation. The chapter highlights a number of research pathways in business value in cloud computing research to guide IS and CS researchers on future avenues of research.

The seven chapters comprising "Measuring the Business Value of Cloud Computing" provide a multidisciplinary perspective on the measurement of business value in the cloud computing context discussing different business value measurement metrics, various cloud service provision models, deployment models, and case studies representing cloud consumers, suppliers, and intermediaries. Cloud computing technology continues to advance in ways that will undoubtedly complicate value measurement. These advances coupled with the dependencies between cloud computing and other advanced technologies such as IoT and Big Data further highlight the need for greater clarity on the definition and appropriate metrics of business value, comprehensive measurement techniques. There is also a need to further untangle the relationships between cloud assets and capabilities, other IS assets and capabilities, and socio-organisation capabilities. To further advance this field of understanding, collaboration between information systems and computer science researchers is both recommended and paramount.

Dublin, Ireland   Theo Lynn
Malibu, CA   John G. Mooney
Dublin, Ireland   Pierangelo Rosati
Dublin, Ireland   Grace Fox

Acknowledgement

This book was partially funded by the Irish Institute of Digital Business at DCU Business School and by the Irish Centre for Cloud Computing and Commerce, an Enterprise Ireland and IDA funded Technology Centre.

Book Description

"Measuring the Business Value of Cloud Computing" explores how the increasingly complex area of cloud computing generate business value and how this value is captured and measured by enterprises. Recent reviews of extant literature highlight the need for multi-disciplinary research that both explores and further develops the conceptualization of value in cloud computing research and investigates how IT value manifests itself across the chain of service provision and in inter-organizational scenarios. This book reviews the state of the art from an information systems, computer science and accounting perspective. It explores and discusses some of the main techniques for measuring the business value of cloud computing in a variety of contexts and illustrates these with mini-case studies. It concludes with futures avenues of research. As such, it provides up-to-date knowledge and methodologies for higher education educators, researchers, students, and industry stakeholders.

CONTENTS

NOTES ON CONTRIBUTORS

Thomas Acton is Head of School of Business & Economics, and a lecturer in Business Information Systems at NUI, Galway, Ireland. His research interests are cloud computing, decision support systems and mobility. He has also served as associate editor on a number of journals, including the European Journal of Information Systems (EJIS) and the Journal of Theoretical and Applied E-Commerce Research (JTAER).

Jörn Altmann is Professor for Technology Management, Economics, and Policy at Seoul National University. Prior to this, he taught at UC Berkeley, worked at Hewlett-Packard Research Labs, and had been a postdoc at EECS and ICSI of UC Berkeley. His research centres on the economic analysis of Internet services and on the integration of economic models into computing systems.

Ram Govinda Aryal received a PhD in Engineering from Seoul National University, South Korea. Currently, he works with the Government of Nepal in the area of technology policy and management. His research interests include cloud federation, economics of cloud services, resource optimization, and technology policy. He has published works in these areas.

Trevor Clohessy is a lecturer in business information systems and transformative technologies at GMIT School of Business. His research interests include blockchain, business analytics, digital transformation, digital addiction, digital politics and cloud computing. Trevor is a member of the GMIT Business school research ethics committee and is a programme coordinator for a blockchain course. Trevor has also been published in leading journals including Information Technology & People and the Journal of Industrial Management & Data Systems.

Marvin Duddek is currently enrolled in the MBA program at Pepperdine University's Graziadio Business School. Born in Germany, he received his Bachelor's Degree in Business Administration from the University of Applied Sciences South Westphalia where he also participated in an apprenticeship program at Siemens. Marvin was ranked among the best IT consultant apprentices in Germany and received a 3-year talent development scholarship from the German Federal Ministry of Education and Research. Prior to enrolling in Pepperdine, he worked as an IT Consultant for Atos SE, conducting strategy workshops for large German-based customers.

Vincent C. Emeakaroha is currently a Lecturer at the Cork Institute of Technology (CIT) Ireland. He received his PhD in Computer Science with excellence from Vienna University of Technology, Austria in 2012. Vincent has authored over 44 international peer reviewed publications. His research interests include Cloud monitoring, Cloud service provisioning management, Cloud trust, Data privacy, cybersecurity, IoT, SLA and QoS.

Grace Fox is an Assistant Professor of Digital Business at Dublin City University Business School. Her research interests intersect the broad interdisciplinary areas of information privacy and digital technology adoption and assimilation. Her research has been published in premier academic journals such as Information Systems Journal and Communications of the Association of Information Systems along with numerous peer-ranked chapters and International conferences in the management, information systems and computer science domains.

Nina Helander is Professor of Knowledge Management. She works at Tampere University as a Head of Unit of Information and Knowledge Management. Her research focuses especially on digitalization, value creation and knowledge management. She has been leading several multidisciplinary research projects including projects that have focused on digitalization and data-based value creation. She is also an Adjunct Professor of Information Systems in University of Jyväskylä.

Theo Lynn is Full Professor of Digital Business at Dublin City University and is Director of the Irish Institute of Digital Business. He was formerly the Principal Investigator (PI) of the Irish Centre for Cloud Computing and Commerce, an Enterprise Ireland/IDA-funded Cloud Computing Technology Centre. Professor Lynn specialises in the role of digital technologies in transforming business processes with a specific focus on cloud computing, social media and data science.

John G. Mooney is Associate Professor of Information Systems and Technology Management and Academic Director of the Executive Doctorate in Business Administration at the Pepperdine Graziadio Business School. Dr. Mooney holds a BS in Computer Science and a Master of Management Science both from University College Dublin, and a Ph.D. in Information Systems from UC Irvine. He was awarded Fellow of the Association for Information Systems in December 2018. His current research interests include management of digital innovation and business executive responsibilities for managing digital platforms and information resources.

Lorraine Morgan is a lecturer in Business Information Systems at NUI, Galway, Ireland. and researcher with Lero: The Irish Software Research Centre, Ireland. Her research has also been published in leading journals including the Journal of Strategic Information Systems, European Journal of Information Systems, Database for Advances in Information Systems and Information and Software Technology.

John Morrison is the founder and director of the Centre for Unified Computing. He is a co-founder and director of the Boole Centre for Research in Informatics and a co-founder and co-director of Grid-Ireland. More recently, he was a principal investigator in the Irish Centre for Cloud Computing and Commerce and Coordinator of the Horizon 2020 CloudLightning project. Professor Morrison has held a Science Foundation of Ireland Principal Investigator award and has written widely in the field of Parallel Distributed and Grid Computing. He has been the guest editor on many journals including the *Journal of SuperComputing, Future Generation Computing Systems and the Journal of Scientific Computing.* He has served on dozens of international conference programme committees and is a co-founder of the International Symposium on Parallel and Distributed Computing.

Arto Ojala is Professor of International Business in the School of Marketing and Communication at University of Vaasa. His research is at cross-section of international business, information systems, and entrepreneurship. He has published in Journal of World Business, Information Systems Journal, Journal of Systems and Software, among others. He is also Adjunct Professor in Software Business at the Tampere University.

Victoria Paulsson is an independent research consultant. Formerly, she was a postdoctoral researcher at Dublin City University in the area of cloud computing. She received PhD in Information Systems from Lund University, Sweden in 2013. Her research interests include accounting information systems, mobile healthcare and cloud computing.

Pierangelo Rosati is Assistant Professor in Business Analytics at DCU Business School and Director of Industry Engagement at the Irish Institute of Digital Business. Dr. Rosati holds a PhD in Accounting and Finance from the University of Chieti-Pescara (Italy) and an MSc in Management and Business Administration from the University of Bologna. He was appointed Visiting Professor at the University of Edinburgh Business School, Universidad de las Américas Puebla and at Católica Porto Business School, and visiting Ph.D. Student at the Capital Markets Cooperative Research Center (CMCRC) in Sydney. Dr. Rosati has been working on research projects on FinTech, Blockchain, cloud computing, data analytics, business value of IT, and cyber security.

Paul P. Tallon is Professor of Information Systems, Law, and Operations at the Sellinger School of Business and Management at Loyola University Maryland. He has published in many of the top journals in the field of Information Systems including MIS Quarterly, the Journal of MIS, the Journal of Strategic Information Systems, the European Journal of Information Systems, Communications of the ACM, the Journal of the Association for Information Systems, and the Journal of Information Technology. He previously worked as a forensic accountant/IT auditor with PricewaterhouseCoopers in Dublin, Ireland and New York, NY.

Pasi Tyrväinen is Professor of information systems and Dean of the Faculty of Information Technology at University of Jyväskylä. His background includes fourteen years at Honeywell and Nokia Research Center. He has 100+ publications on IS themes including enterprise content management, software business and others in EJIS, IST, JIM, JSS, DSS, IEEE Software and others.

ABBREVIATIONS

AI	Artificial Intelligence
AIOps	AI for IT Operations
API	Application Programming Interface
AWS	Amazon Web Services
B2B	Business To Business
B2C	Business To Consumer
BPAAS	Business Process As A Service
BVCC	Business Value of Cloud Computing
BVI	Business Value Index
C2T	Cloud-to-Thing
CAPEX	Capital Expenditure
CFO	Chief Finance Officer
CIO	Chief Information Officer
CISR	Centre for Information Systems Research
CPU	Central Processing Unity
CRM	Customer Relationship Management
CS	Computer Science
CSB	Cloud Service Broker
CSF	Critical Success Factor
EC2	Elastic Compute Cloud
EMEA	Europe, Middle East and Africa
EVA	Economic Value Added
FAAS	Function As A Service
FEDRAMP	Federal Risk And Authorisation Management Program
FPGA	Field-programmable Gate Array
FSLA	Federation Service Level Agreement
FX	Foreign Exchange

GPS	Global Positioning System
GPU	Graphics Processing Unit
HDD	Hard Disk Drive
HP	Hewlett Packard
HPC	High Performance Computing
HVAC	Heating, Ventilation, and Air Conditioning
IAAS	Infrastructure As A Service
IOT	Internet Of Things
IPTV	Internet Protocol Television
IRR	Internal Rate of Return
IS	Information Systems
ISACA	Information Systems Audit and Control Association
ISO	International Organization for Standardization
IT	Information Technology
ITSP	Information Technology Service Provider
KPI	Key Performance Indicator
MAAS	Marketplace As A Service
MIT	Massachusetts Institute of Technology
ML	Machine Learning
NIST	National Institute of Standards and Technology
NPV	Net Present Value
OPEX	Operating Expenditure
OS	Operating System
PAAS	Platform As A Service
PC	Personal Computer
PCI DSS	Payment Card Industry Data Security Standard
POS	Point Of Sale
QoE	Quality of Experience
QoS	Quality of Service
RAM	Random Access Memory
RFID	Radio Frequency Identification
RFX	Retail Foreign Exchange
ROI	Return On Investment
SAAS	Software As A Service
SIM	Society For Information Management
SLA	Service Level Agreement
SQL	Structured Query Language
SSO	Single Sign On
STOF	Service, Technology, Organisation and Finance
TCO	Total Cost Of Ownership
VM	Virtual Machine

LIST OF FIGURES

LIST OF TABLES

CHAPTER 1

Measuring the Business Value of IT

Paul P. Tallon, John G. Mooney, and Marvin Duddek

Abstract Firms have consistently struggled to measure the business value of information technology (IT). In an era where IT is transitioning to a services model, firms are replacing capital expenditure with operating expenditure. The implications for IT business value measurement are significant. In this chapter, we examine the state of knowledge about IT business value with particular emphasis on established metrics for IT business value. We then consider how these metrics might be applied to cloud-based services. The move to a services model further provides an opportunity to consider IT business value in a new light by considering how cloud technologies enhance IT agility, how firms can monetise their data, and how firms now have greater flexibility around IT use than ever before.

P. P. Tallon (✉)
Loyola University Maryland, Baltimore, MD, USA
e-mail: pptallon@loyola.edu

J. G. Mooney • M. Duddek
Graziadio Business School, Pepperdine University, Malibu, CA, USA

T. Lynn et al. (eds.), *Measuring the Business Value of Cloud Computing*, Palgrave Studies in Digital Business & Enabling Technologies, https://doi.org/10.1007/978-3-030-43198-3_1

1

Keywords Return on IT investment • Cloud technologies • Net present value • IT performance

1.1 INTRODUCTION

In an era defined by rapid adoption of information technology (IT) services and increasing volumes of data, senior IT executives continue to emphasise the need for measuring and managing the business value of IT investments (Kappelman et al. 2019). Annual surveys of top IT executives by the Society for Information Management (SIM) attest to significant growth in IT spending as a percentage of revenues with IT spending growing from 3.8% of revenues in 2008 to 5.9% in 2018 (Kappelman et al. 2019). The composition of IT spending has also shifted over time, reflecting the adoption of cloud-based IT services and the associated shift from viewing IT investment as capital expenditure (CapEx) to operating expenditure (OpEx). While data analytics and cybersecurity are among the largest categories of IT spending, spending on cloud computing (Software-as-a-Service—SaaS, Platform-as-a-Service—PaaS, and Infrastructure-as-a-Service—IaaS) is growing and expected to outpace other areas of expenditure in the coming years (Kappelman et al. 2019). This shift will create opportunities for CIOs to rethink how they allocate IT resources. It also puts pressure on CIOs to think about how they should justify and manage new areas of IT spending. In the past, IT projects were required to undergo a significant and often time-consuming budget approvals process. However, the move to cloud computing means that IT investment is no longer *up-front* but rather distributed over time based on resource utilisation. The net result is that the determination of IT business value has become less of a preoccupation for CIOs and business executives when IT spending—like a subscription for cloud-based services—is regarded as OpEx rather than CapEx. Yet, the overarching question still remains: how can CIOs assess the business value of IT when IT is provisioned and utilised as a service in the cloud? In this chapter, we explore this question by looking beyond current IT business value measurement to ways of evaluating the unique benefits that flow from cloud-based IT services. Much of the benefits of cloud computing, we propose, are found in the ability of organisations to use cloud computing to scale and adapt IT services and to generate process agility through systems development, deployment, and retirement, thereby providing a greater range of IT support and flexibility to critical business functions.

1.2   Financial Approaches to Measuring Business Value

Managers have long been advised to think about IT investments through the lens of capital budgeting: an IT investment calls for an initial capital outlay which is followed in later time periods by a stream of predictable benefits that can be modelled as incremental cash flows. Discounting—adjusting for the time value of money—can then be applied to produce a true measure of value. In Table 1.1, we summarise standard metrics that IT executives have long considered necessary to help managers compute and articulate the business value of IT. Yet, as Tallon and Kraemer (2006) note, the often intangible nature of some IT benefits such as enhanced customer satisfaction or improved employee morale present intractable cash flow measurement challenges, making it progressively difficult for CIOs and their business partners to accurately determine IT business value.

Recognising the complexity of expressing various IT impacts in terms of incremental cash flows, researchers have turned to alternative metrics to get closer to the actual impacts themselves. These researchers suggest using process-level perceptual measures of business value rather than firm-level financial measures. Tallon and Kraemer (2007), for example, conclude that perceptual measures are a valid proxy for non-existent or hard-to-find financial measures of IT business value such as those in Table 1.1. Thus, as we turn to the question of how to compute IT business value in an era of cloud services, we can apply findings from the extant literature to propose that perceptual measures are a reasonable way to consider IT business value and that measuring IT business value at the process-level is preferable to pursuing more aggregated firm-level financial measures.

1.2.1   OpEx Measures of IT Business Value

Unlike traditional IT investments that are ordinarily regarded as CapEx, spending on cloud-based services using subscription or usage-based payments are treated as OpEx. As such, payments are directly expensed in the income statement in the time period in which the service is consumed. Unless the user has purchased physical hardware or software which is hosted by the cloud provider, the user does not own any IT assets. Absent an IT asset, nothing appears on the balance sheet. This fundamentally alters the conversation around the use of NPV, ROI, IRR, EVA or payback

Table 1.1 Standard IT investment evaluation metrics

	Description of metric
Net Present Value (NPV)	NPV is one of the foremost financial key performance indicators (KPIs) used to evaluate large, capital-intensive IT projects. NPV relies on accurate cash flow projections extending over the life of the project, alongside a discount rate which is used to account for the time value of money. Project approval depends on obtaining a positive NPV. IT projects can also be compared with one another using NPV whenever firms need to ration scarce IT capital.
Return on Investment (ROI)	ROI is an accounting-based ratio that compares total project income to the level of project investment. ROI does not take account of the time value of money, meaning that projects with a longer-term return window would be treated on par with projects that generate equal returns over a shorter time period. Similar to NPV, the accuracy of ROI calculations depends on being able to identify the scale of future cash flows arising from an investment.
Payback period	The payback period is a simplistic method that calculates the time needed for a project to breakeven (recover its investment costs). In a risk averse firm, managers may gravitate towards IT projects with a shorter payback period. In practice, payback should not be used in isolation but rather alongside other metrics that take account of project risk and that consider the flow of benefits beyond the end of the payback period.
Internal Rate of Return (IRR)	Given all future cash flows and an upfront investment for an IT project, IRR is the discount rate that would return a value of zero for NPV. IRR can be considered the true rate of return in that it takes account of the time value of money and the flow of value over time. IRR can be benchmarked against desired or minimum rates of return, including the weighted cost of capital.
Economic Value Added (EVA)	EVA—also called economic profit—is a measure of residual value generated by a project after deducting the cost of invested capital. Since all capital can be allocated to different ends, EVA argues that projects should be assessed an investment cost. This allows for a more equitable comparison if managers are in a position to pick from different IT projects with unique rates of return.
Total Cost of Ownership (TCO)	TCO captures a multitude of different cost items in a single metric such as the cost of hardware, software, and services, allocated per application, user, department, etc. TCO can also be represented as a cost per period of time. TCO does not take into account the benefit or value to the organisation of using the underlying resource and is, as such, a questionable metric unless accompanied by other metrics such as ROI, NPV or payback period.

Source: adapted from Keen and Digrius (2003)

metrics since there is no initial capital outlay in the way that we normally expect an upfront outlay to apply these traditional measures. Any initial outlay is replaced by a stream of payments over time. Unless these payments are predictable as in the case of a fixed subscription price model, it can be difficult to use discounting (as one would with NPV or IRR) to obtain a clear view of the real cash outflows.

In order to fully appreciate why these metrics are problematic in a cloud setting, it is useful to understand why IT projects have traditionally been characterised by large upfront capital outlays that are then depreciated to zero over multiple periods of time. Before the availability of large scale IT resource sharing, organisations had to acquire and own their IT resources. Once the demand for IT resources was established, that IT resource demand could be met with dedicated assets. The rise of server sprawl was an unfortunate albeit predictable side effect of this process if requests for new applications, for example, led to the creation of dedicated servers (McFarlan and Bartlett 2002). If security, trust and data privacy were concerns, it was often seen as better for firms to use dedicated internal IT resources rather than using shared IT resources, either internally with other departments or employees or externally with third parties. Provisioning of dedicated internal IT resources was further complicated by the need for these resources to satisfy peak demand. Since the level of IT resources could not scale easily or quickly, anticipated spikes in demand for IT resources meant that firms had to raise their initial IT investment to the level necessary to satisfy the projected peak, and then carry that excess capacity (and its associated costs) through periods of reduced demand. The net effect of this approach to resource management was to dilute the level of IT business value since some portion of the IT resources lay idle for significant periods yet these resources had to be purchased, managed and would continue to depreciate in value over time, regardless of actual use.

Cloud-based services with their focus on OpEx rather than CapEx transform this dynamic of provisioning IT resources. When firms are faced with the question of whether to build or buy IT functionality, it is increasingly likely that they will buy standardised IT resources off-the-shelf in ways that they can quickly scale usage of the resource. Effectively, the rise of cloud-based "IT-as-services" means that the supply of IT resources should always equal the actual demand for IT. From the perspective of alignment or fit between IT and business strategy—a perennial concern for CIOs (Kappelman et al. 2019)—the availability of *on-demand* cloud-based services means that firms' IT resources are more likely to be persistently aligned with business needs than if they were to build and manage internal IT resources with their attendant delays in responding to IT demands.

Since classic CapEx-based metrics are unable to capture the full portfolio of benefits from cloud-based solutions, CIOs must search for alternative ways to evaluate IT projects. In so doing, they have an opportunity to look outside the strict boundaries of incremental cash flows. There may be intangible benefits—often ignored in NPV and IRR calculations—which can be as valuable and as desirable as tangible benefits. In order to give a full accounting of IT business value in a cloud-based IT services environment, we must be able to correctly identify such impacts, notwithstanding the difficulty of finding or measuring them (Hares and Royle 1994; Keen and Digrius 2003). Some of these qualitative IT benefits are sought in traditional IT investments also, notably improvements in customer satisfaction, employee morale or product and service quality (Hares and Royle 1994, p. 206). When identifying IT business value derived from the utilisation of cloud computing, there are some very specific impacts that can be evaluated, notably:

1. Resilience: a risk-based measure that speaks to system reliability and availability.
2. Speed of Deployment: since deployment tends to lag any decision to deploy IT resources, this measure assesses the ability of IT to respond to changes in the demand for IT.
3. Scalability: describes how easily and quickly incremental IT resources can be added to (or removed from) the portfolio of IT resources available to distributed users.
4. Organisational Agility: describes how easily and quickly organisations can respond to changes in their business environment and, more importantly, at what cost.

The above metrics provide a window into a range of non-financial impacts. Unfortunately, it can be difficult to quantify these impacts and so the risk of mismeasurement remains. Indeed, it is also possible that firms could pursue cloud-based solutions not because of immediate financial considerations (cost savings, for example) but rather because they see the cloud as enabling them to do things that would otherwise not be possible. Moving to the adoption of cloud computing also opens up possibilities for firms to focus more on effective implementation and use of systems and information—the I in IT—rather than continuing to attribute value to ownership of hardware and software—the T in IT. In the next section, we offer a brief example and some detailed discussion on what this might mean for organisations.

1.3   BEYOND FINANCIALS: QUANTIFYING NON-FINANCIAL ASPECTS OF IT BUSINESS VALUE

Although the IT business value literature argues that process-oriented perceptual measures are an acceptable proxy for unavailable or difficult-to-find financial measures, managers need to recognise that perceptions are still subject to individual error and bias. Hence, Tallon (2014) argues that IT decision makers may need to rely more on the views of multiple IT and business executives at different levels within the organisation in what he describes as a *distributed sensemaking* model. Even if one executive has a less than perfect view of how IT is performing, their biased views can be balanced by insights gleaned from other executives within the organisation. Tallon (2014) notes that a key to enabling business executives to improve their ability to perceive IT business value is having IT executives engage in *sensegiving* activities. CIOs and the information systems function can, in an educative role, assist their business peers in ways to think about IT business value. They are not telling their business partners *what to think* about IT but rather *how to think* about IT. One such exercise involves ways to convert intangible views of IT impacts into more tangible outcomes (Hares and Royle 1994, p. 206). We review two such methods below: Scoring and Value Linking.

1.3.1   *Scoring*

Scoring seeks to assign weights and values to different outcome criteria, making it possible to prospectively analyse and compare IT performance under different scenarios (Keen and Digrius 2003). In Table 1.2, we provide a brief demonstration of what this might look like when comparing a cloud-based IT solution to an on-premise solution, using a hypothetical set of weighted decision criteria. Scores are simply the product of weights multiplied by an estimated grade on a scale from −5 to +5. Grades represent desirable and undesirable outcomes. In the case of undesirable

Table 1.2 An illustrative example of how scoring can be used to compare IT solutions

Decision criteria	Weights	Cloud-based solution		On-premise solution	
		Grade (−5 to 5)	Score	Grade (−5 to 5)	Score
Enhanced user experience	15	4	60	4	60
Risk of user acceptance	10	−1	−10	−1	−10
Scalability	20	5	100	2	40
Failover scenario	15	5	75	3	45
Level of access to information	10	5	50	3	30
Security infrastructure	20	5	100	3	60
Risk of storing data externally	10	−3	−30	5	50
Total	100		345		275

Source: adapted from Keen and Digrius (2003, p. 126)

outcomes such as a rise in risk, incrementally undesirable outcomes can be modelled using negative scores.

Although the scoring method is relatively easy to apply, the main challenge is to cope with each individual's subjectivity. In order to overcome the potential for bias, multiple individuals can be asked what they consider to be suitable weights and scores. However, the group must first agree on the range of intangible impacts; otherwise, some key impacts could be missed. Once the overall structure is agreed, those involved in the process may submit their values and an overall weighted score can be determined. Interrater reliability scores could then be used to ascertain the degree of consistency in the group. Furthermore, it would be beneficial if individuals were allowed to discuss their weights and scores so that opposing viewpoints can be identified and reconciled (Buss 1983).

1.3.2   Value Linking and Value Acceleration

Value Linking and Value Acceleration are related concepts that aim to financially evaluate how the initial intangible benefits attributed to IT trickle down to the financial performance of the organisation. Whereas Value Linking focuses on the effects that IT has on factors such as revenue and cost, Value Acceleration aims to identify the range of one-off or unique IT impacts (Parker et al. 1988). In order to examine the second-order or firm-level effects of IT, it is necessary to identify the range of first-order effects at critical points within the value chain. Different technologies will impact processes such as supplier relations or product and service innovation in different ways. So, the secondary effects of IT on revenues, for example, can be tied to a host of first-order benefits that tie incremental revenues to enhanced IT-led supplier relations or to enhanced IT-led product or service innovation. A useful starting point is to trace IT impacts back from five broad categories noted by Sassone and Schwartz (1986): operations savings, performance improvements, increased sales, labour savings, and shorter conversion cycles or to critical business processes within the value chain such as supplier relations, customer relations, product and service enhancement, production and operations, and sales and marketing support (Porter 1985; Tallon 2008; Tallon et al. 2000).

1.4   BEYOND QUANTIFICATION: HOLISTIC MEASUREMENT OF IT BUSINESS VALUE

In the end, the determination of IT business value is part *art* and part *science*. The inclusion of subjective measurement can never be fully discounted since the causal paths along which IT generates value for organisations are crisscrossed with any number of factors that can confuse the true relationship between IT and business performance. Recognising this, the move to cloud-based IT services suggests that managers may want to consider more holistic measures than simply trying to link IT to financial performance. One way to develop this holistic view of IT business value is to recognise that cloud-based IT solutions generate digital options that allow organisations to scale or to change direction much faster than in cases where the firm owns and runs its own IT resources inside its owned and operated data centre (Sambamurthy 2000; Sambamurthy et al. 2003). Moving to cloud computing also allows us to recognise that at a time when IT services are becoming more standardised, effective application of those services to better manage data and business processes may emerge as sources of competitive differentiation. Any move to recognise data as a strategic asset that can be placed on the balance sheet has been controversial. Yet the literature is beginning to make the case that data should be monetised but that it also requires novel forms of governance (Buff et al. 2015; Short et al. 2011; Tallon et al. 2014). Lastly, one could consider implementing a Business Value Index (BVI), as illustrated by Intel (Curley 2003). BVI considers IT business value on a two-dimensional grid. The first dimension—IT Efficiency—asks how well the proposed IT investment utilises Intel's established IT infrastructure. The second dimension—Business Value—considers the impact of the proposed IT investment on Intel's business. Investments can be scored as low (−1), medium (0) or high (+1) on each dimension, meaning that investments can be assigned to any one of nine possible positions inside a 3x3 grid. Potential investments that score low on one or both dimensions are unlikely to be funded, whereas those rated high on one or both dimensions stand a greater chance of being funded. Potential investments that fall between these two extremes can be postponed to a later period in the hopes that IT efficiency and business value might improve, abandoned entirely or could be funded in whole or in part based on the availability of IT resources and the firm's risk propensity (Fichman et al. 2006; Tallon et al. 2002).

1.4.1   Digital Options in the Cloud

The challenge with assessing IT business value using a capital budgeting framework is the need to create a model in which benefits and costs are expressed as positive or negative cash flows which are, in turn, restated according to the time value of money. *Risky* projects are discounted at a higher rate, making it far less likely they will be funded (Kambil et al. 1993; Tallon et al. 2002). If these same IT investments were examined through the lenses of digital options, however, it is entirely possible that an IT investment with a negative net present value could still have a positive options value and be funded on that basis. Moving IT to the cloud provides firms with a variety of options which are often overlooked when using a capital budgeting framework: the option to easily and quickly scale an investment, the option to delay a time-sensitive investment, the option to fund a small-scale pilot project, the option to cancel an investment without penalty or sunk cost, and the option to build out an IT investment in stages (Fichman et al. 2006). As such, an options valuation framework may be a better way to think about creating and managing cloud-based IT investments.

1.4.2   Monetisation of Big Data

The discovery during bankruptcy proceedings in 2015 that gaming data housed by Caesars Entertainment was valued at more than $1 billion highlights the potential value of data assets that organisations have yet to include in the assets section of their balance sheets (Marr 2015). Not all data assets are equally valuable and yet for a growing number of organisations, data is increasingly seen as a strategic asset. The cost of creating or acquiring data is viewed as an operational expense, meaning that the expense is written off in the year in which it is incurred; hence, no portion of the cost is capitalised as an asset for inclusion on the balance sheet. Despite the fact that data might be worth a multiple of what it costs to create, purchase, store, and manage on an ongoing basis, there is no formal mechanism for recognising data as an asset or for determining how data assets should be valued on a continuous basis (Tallon and Scannell 2007). Moving data to the cloud, potentially affords a way to classify data according to its usefulness. Data classification attempts to place data within a lifecycle curve that extends from the moment of its creation (or acquisition) to the moment of its death (or deletion). If cloud storage is priced according to frequency of access, it is possible to isolate data that is frequently used and likely of higher value to the organisation and data that is less frequently accessed and, therefore, of lower value to the organisation.

Besides the question of how data can be valued, there is also the question of how data can be used to create value. The literature has previously spoken of the effects of IT without separating out the incremental effects of technology (hardware, software, and telecoms) from those of data or information. That may be about to change with IT transitioning to the cloud. Research from MIT's Center for Information Systems Research (CISR) describes how data can be monetised by selling, exchange/bartering, and wrapping. In the case of selling, data can be sold to the highest bidder. In the case of exchange/bartering, data may be exchanged for something of equal value. Retailers, for example, may be willing to share data with third parties who, by aggregating data from multiple retailers, are able to uncover insights that may not be discoverable by any one retailer in isolation. Data wrapping meanwhile describes how some organisations such as Fitbit can add data and data analytics features or capabilities to their products or services (Wixom and Schüritz 2018). The use of cloud computing amplifies an organisation's ability to do this given the need for massive storage and computing capabilities to generate customised user experiences with data. What this means is that cloud technologies have made it possible for organisations to monetise data in ways that might not be otherwise possible were data to be locked in silos deep within company-owned data centres.

1.5   ADDITIONAL CONSIDERATIONS FOR BUSINESS VALUE OF CLOUD COMPUTING

The fact that cloud-based IT resources are now regarded as OpEx rather than CapEx does not mean that the CapEx versus OpEx debate has been resolved once and for all. Companies that have moved to an OpEx or services model need to be careful because of how pricing relates to IT resource utilisation. In the same way that apartment leasing in perpetuity can be far more expensive than home ownership, it is possible that cloud costs could, if left unchecked, total to more than if the IT assets were owned and managed by the enterprise itself. It is not sufficient to justify moving to the cloud at a point in time and to then ignore cloud-based costs from that time forward. Whereas the cost of an owned IT asset is a sunk cost—you cannot recoup your entire investment cost if you decide that an IT project is no longer viable—cloud costs are ongoing. What this means is that IT managers will likely need to repeatedly justify their decision to use the cloud. They have the option to bring IT back in-house if costs rise. If the total cost of owning IT drops to where cloud solutions are no longer economically attractive, the CapEx versus OpEx debate could reignite once more.

This discussion is especially relevant in the context of a move by some companies to create platform technologies. At a time when organisations are increasingly structured by business units with geographic, customer, product or market oversight, corporate-managed IT platforms are seen as a way to support certain processes with shared IT resources. The challenge is that some business units might push back against the idea of using shared IT resources to support an activity that they feel is sufficiently unique to warrant direct support from the business unit itself. As argued by Ross et al. (2006), few organisations have mastered the task of building an IT platform that meets both shared and idiosyncratic needs. Dealing with the tension between business units and their corporate parent can mask the same trade-offs between CapEx and OpEx if there is a decision to allow some units to invest in their own IT (CapEx) while others are supported by shared resources in the cloud (OpEx). Here again, managers must question whether standardised IT resources are unable to meet idiosyncratic IT needs as some business units believe or whether *simple IT*, in the words of Upton and Staats (2008), can meet all IT needs. If all needs can be met through standardised, cloud-based IT, the argument then follows that IT platforms should be implemented using cloud technologies.

Of course, any move to control business units' IT choices by supporting their IT needs via a shared, corporate-controlled IT platform fails to account for the inevitability of shadow IT. At a time when users can use the cloud to meet any immediate IT resource needs—circumventing IT policies or governance systems that might otherwise cause delays—the lure of cloud-based shadow IT is understandable. As noted by Intel's former Chief Privacy Officer, Malcolm Harkins, the point is not to lock down the use of shadow IT for reasons of risk management or for cost avoidance but rather to create cloud-friendly policies that recognise the value of the cloud when users are pressed for time or resources (Harkins 2013). As long as shadow IT exists within a governance framework that sees the value of allowing users to self-serve when IT platforms are insufficient or when an IT solution must be identified expeditiously, the value of using scalable cloud technologies can be significantly greater than if IT impacts are measured solely through the eyes of the corporate entity.

1.6 Beyond Measurement to Managing Benefits Realisation

It is one thing to measure IT business value with any degree of accuracy but it is another thing to manage IT in a way that maximises the potential for value creation. IT business value does not emerge automatically. Lee Iacocca, President, CEO, and Chairman of Chrysler from 1978 to 1992, famously claimed that he had approved so many IT projects involving employee layoffs that there should be nobody working at Chrysler (Iacocca 1989, p. 243). IT investments are no different in that the promised benefits of IT will only be realised if IT is effectively managed. This means that failed or failing projects may need to be cancelled or redirected if there is a possibility that the expected benefits will not be realised. However, as Keil (1995) observed, cancelling an apparently failing IT project is easier said than done. IT managers are more likely to drag out failing projects long after they should have been paused, cancelled or reviewed. The lesson in this carries over from our prior discussion about moving from CapEx to OpEx. Despite the attractiveness or convenience of cloud technology, there is no point in continuing a subscription to an IT service that has outlived its usefulness or if there is greater value if the underlying technology is brought back in-house.

Value realisation also means that there should be some form of accountability for effective use of cloud services, since without effective use it will be impossible to realise IT business value (Devaraj and Kohli 2003). Not all forms of IT use will be value generating, but if IT is not used or is not useful, business value will simply fail to arise. It makes sense, therefore, for organisations to use periodic reviews to ascertain if IT resources and services are living up to their promise. So called post-implementation or after-action reviews help managers to discover what works and what doesn't. The reality, however, is that fewer than 30% of organisations perform *any* form of post-implementation review once a CapEx investment is live (Tallon et al. 2000). For cloud technology, it is likely to be considerably lower since cost is treated as OpEx and is almost certain to be below the minimum threshold of what most firms need to trigger or justify a post-implementation review. From a cognitive perspective, this means that it may be harder for managers to focus on whether the business value from cloud technologies is appropriate since they may never have to go through any formal review to illustrate the level and locus of IT business value. To avoid such a situation, managers may need to monitor the level and frequency of cloud use and undertake reviews of user satisfaction. User satisfaction is a proxy for business value only insofar as it helps to detect gaps between what users want from IT and what IT does for them.

Lastly, there is a question of whether use of cloud technology creates a risk of over reliance on third-party providers and whether there may be an erosion of critical IT skills from within the organisation. Just as IT outsourcing led some firms to lose key IT skills (Austin et al. 2005), there is a risk that any move to bring IT applications or critical IT infrastructure back in-house might be impeded by a lack of internal IT skills. Such a move could have a detrimental effect on IT benefits in the long term and could lock firms into using less-than-adequate (and expensive) IT solutions.

1.7   Conclusion

The movement of IT applications and resources from owned, on-premise assets to third-party cloud-based services has had a profound impact on organisations' ability to use IT to support their business functions. Beyond the benefits of reach and scale and associated impacts on IT risk mitigation and IT support cost, the ability to treat cloud expenses as OpEx rather than CapEx has changed the way that managers think about IT business value. From a time when all large capital IT investments were required to undergo a formal cost-benefit assessment prior to acquisition, it is now possible for organisations to avoid such analysis since IT costs are now largely based on IT utilisation and the expense is often believed to be sufficiently low to negate the value of a required cost-benefit analysis. It would be foolhardy for organisations to adopt this view in perpetuity since early indications are that over an extended timeframe cloud-based IT solutions can be as expensive if not more expensive than their on-premise equivalent. There is, as such, a need to think carefully about IT business value in the context of cloud computing. It is not simply a matter of where IT support originates—from the cloud or from a company-owned data centre; IT business value is not the same either way. Transitioning to the cloud uncovers additional types of value (options around process agility, for example) that might not be immediately apparent in on-premise situations, but such a move also triggers the potential for other costs to arise. As the late Andy Grove noted about measurement at Intel, you cannot manage what you cannot measure (Grove 1999). In that regard, cloud computing is no different from any other organisational resource.

References

Austin, R., W. Ritchie, and G. Garrett. 2005. *Volkswagen of America: Managing IT Priorities*. Harvard Business School Press. [No. 9-606-003].

Buff, A., B.H. Wixom, and P.P. Tallon. 2015. *Foundations for Data Monetization*. MIT Sloan CISR Working Paper No. 402.

Buss, M.D.J. 1983. How to Rank Computer Projects. *Harvard Business Review* 61 (1): 118–125.

Curley, M. 2003. *Managing Information Technology for Business Value: Practical Strategies for IT and Business Managers*. Hillsboro, OR: Intel Press.

Devaraj, S., and R. Kohli. 2003. Performance Impacts of Information Technology: Is Actual Usage the Missing Link? *Management Science* 49 (3): 273–289.

Fichman, R.G., M. Keil, and A. Tiwana. 2006. Beyond Valuation: Options Thinking in IT Project Management. *California Management Review* 47 (2): 74–96.

Grove, A.S. 1999. *Only the Paranoid Survive: How to Exploit the Crisis Points That Challenge Every Company*. New York, NY: Doubleday.

Hares, J.S., and D. Royle. 1994. *Measuring the Value of Information Technology*. New York, NY: John Wiley.

Harkins, M. 2013. *Managing Risk and Information Security: Protect to Enable*. New York, NY: Apress Media, LLC.

Iacocca, L. 1989. *Talking Straight*. New York, NY: Bantam Books.

Kambil, A., J. Henderson, and H. Mohsenzadeh. 1993. Strategic Management of Information Technology Investments: An Options Perspective. In: R.D. Banker, R.J. Kauffman, M.A. Mahmood (Eds.), *Strategic Information Technology Management*. Idea Group Publishing: Harrisburg, PA, pp. 161–178.

Kappelman, L., R. Torres, E. McLean, C. Maurer, V. Johnson, and K. Kim. 2019. The 2018 SIM IT Issues and Trends Study. *MIS Quarterly Executive* 18 (1): 51–84.

Keen, J.M., and B. Digrius. 2003. *Making Technology Investments Profitable: ROI Road Map to Better Business Cases*. Hoboken, NJ: Wiley & Sons.

Keil, M. 1995. Pulling the Plug: Software Project Management and the Problem of Project Escalation. *MIS Quarterly* 19 (4): 421–447.

Marr, B. 2015. Big Data at Caesars Entertainment—A One Billion Dollar Asset?. Accessed 20 May 2019. https://www.forbes.com/sites/bernard-marr/2015/05/18/when-big-data-becomes-your-most-valuable-asset/#53eec7131eef.

McFarlan, F.W., and N. Bartlett. 2002. *Postgirot Bank and Provment AB: Managing the Cost of IT Operations*. Harvard Business School. [Case 302-061].

Parker, M.M., R.J. Benson, and H.E. Trainor. 1988. *Information Economics: Linking Business Performance to Information Technology*. Englewood Cliffs, NJ: Prentice Hall.

Porter, M.E. 1985. *Competitive Advantage*. New York, NY: Free Press.

Ross, J.W., P. Weill, and D.C. Robertson. 2006. *Enterprise Architecture as Strategy*. Cambridge, MA: Harvard Business School Press.

Sambamurthy, V. 2000. Business Strategy in Hyper-Competitive Environments: Rethinking the Logic of IT Differentiation. In *Framing the Domains of IT Management: Projecting the Future Through the Past*, ed. R. Zmud, 245–261. Cincinnati, OH: Pinnaflex.

Sambamurthy, V., A. Bharadwaj, and V. Grover. 2003. Shaping Agility Through Digital Options: Reconceptualizing the Role of Information Technology in Contemporary Firms. *MIS Quarterly* 27 (2): 237–263.

Sassone, P.G., and A.P. Schwartz. 1986. Office Information Systems Cost Justification. *IEEE Aerospace and Electronic Systems Magazine* 1 (8): 21–26.

Short, J.E., R.E. Bohn, and C. Baru. 2011. *How Much Information 2010? Report on Enterprise Server Information*. Global Information Industry Center, UC San Diego.

Tallon, P.P. 2008. A Process-Oriented Perspective on the Alignment of Information Technology and Business Strategy. *Journal of Management Information Systems* 24 (3): 231–272.

———. 2014. Do You See What I See? The Search for Consensus Among Executives' Perceptions of IT Business Value. *European Journal of Information Systems* 23 (3): 306–325.

Tallon, P.P., and K.L. Kraemer. 2006. The Development and Application of a Process-Oriented Thermometer of IT Business Value. *Communications of the AIS* 17 (45): 1–52.

———. 2007. Fact or Fiction? A Sensemaking Perspective on the Reality Behind Executives' Perceptions of IT. *Journal of Management Information Systems* 24 (1): 13–54.

Tallon, P.P., and R. Scannell. 2007. Information Lifecycle Management. *Communications of the ACM* 50 (11): 65–69.

Tallon, P.P., K.L. Kraemer, and V. Gurbaxani. 2000. Executives' Perceptions of the Business Value of Information Technology: A Process-Oriented Approach. *Journal of Management Information Systems* 16 (4): 145–173.

Tallon, P.P., R.J. Kauffman, H.C. Lucas, A. Whinston, and K. Zhu. 2002. Using Real Options Analysis for Evaluating Uncertain Investments in Information Technology: Insights from the ICIS 2001 Debate. *Communications of the Association for Information Systems* 9: 136–167.

Tallon, P.P., R.V. Ramirez, and J.E. Short. 2014. The Information Artifact in IT Governance: Towards a Theory of Information Governance. *Journal of Management Information Systems* 30 (3): 141–177.

Upton, D.M., and B. Staats. 2008. Radically Simple IT. *Harvard Business Review* 86 (3): 118–124.

Wixom, B.H., and R. Schüritz. 2018. *Making Money from Data Wrapping: Insights from Product Managers (Research Briefing).* MIT Center for Information Systems Research.

CHAPTER 2

Measuring the Business Value of Infrastructure Migration to the Cloud

Pierangelo Rosati and Theo Lynn

Abstract Infrastructure-as-a-Service (IaaS) and Platform-as-a-Service (PaaS) adoption typically require radical changes in an organisation's IT operations and have widespread implications that go beyond simple cost savings. This chapter presents a practical framework for estimating the Return on Investment (ROI) for IaaS and PaaS from the customer perspective. The proposed framework aims to overcome the main limitations of commonly-used Total Cost of Ownership (TCO) calculators by including both tangible and intangible costs and benefits to provide a more comprehensive ROI estimation. The application of the framework is illustrated using a real-life case study of infrastructure migration.

Keywords Infrastructure-as-a-service • Platform-as-a-service • Return-on-investment

P. Rosati (✉) • T. Lynn
Irish Institute of Digital Business, DCU Business School, Dublin, Ireland
e-mail: pierangelo.rosati@dcu.ie

© The Author(s) 2020
T. Lynn et al. (eds.), *Measuring the Business Value of Cloud Computing*, Palgrave Studies in Digital Business & Enabling Technologies, https://doi.org/10.1007/978-3-030-43198-3_2

19

2.1   Introduction

Cloud computing platforms and applications are proliferating across firms of all sizes worldwide becoming the *de facto* computing paradigm of choice. According to IDG (2018), 73 percent of organisations have at least a portion of their computing infrastructure already in the cloud, and another 17 percent plan to adopt cloud solutions within the short-term. While cloud computing is a well-known reality for large enterprises today, recent years have seen a surge in cloud spending by smaller organisations (IDG 2018). This has resulted in significant growth in the public cloud services market which is now projected to reach $331.2 billion by 2022 (Gartner 2019). Software-as-a-Service (SaaS) is the most common type of cloud computing service and accounts for approximately 43 percent of the market while infrastructure-related services i.e. IaaS and PaaS account for approximately 25 percent of the current market and are experiencing the highest growth rate (Gartner 2019).

The technical benefits of the cloud are well-documented and typically relate to on-demand, self-service resources orchestration, resource pooling and elasticity (Armbrust et al. 2010; Cegielski et al. 2012; Brender and Markov 2013). Cloud computing is also very attractive from a business point of view as it requires lower upfront investment, reduced risk, and improved organisational agility and efficiency (Armbrust et al. 2010; Marston et al. 2011; Leimbach et al. 2014). However, the adoption of cloud computing may also create challenges for firms when an in-depth financial and technical analysis has not been carried out in advance. While selecting the right cloud architecture and the right provider is crucial for an effective delivery of a cloud application, a proper financial analysis is required to make sure the application delivery is sustainable and cost-effective.

As outlined in Chap. 1, there a number a number of methodologies for an *ex-ante* estimation of the business value of cloud migration or adoption (see also Farbey et al. (1993) and Farbey and Finkelstein (2001) for a more detailed discussion) that can be directly applied to IaaS and PaaS services and should be leveraged to better inform the investment decision-making process (Ronchi et al. 2010; Rosati et al. 2019). Total Cost of Ownership (TCO) is arguably the most frequently used technique when it comes to evaluating different cloud vendors and services (Strebel and Stage 2010; Rosati et al. 2017). However, it is important to highlight that

TCO only focuses on cost savings and omits other potential benefits. In contrast, a Return on Investment (ROI) analysis considers the wider strategic implications of cloud adoption and therefore provides a more robust basis for investment decisions (Strebel and Stage 2010; Rosati et al. 2017).

The main objectives of this chapter are to present a practical framework for estimating the ROI on cloud infrastructure and to present a case study to demonstrate how the framework can be applied to an infrastructure migration scenario. The reminder of this chapter is organised as follows. Next, we provide an overview of IaaS and PaaS. Then we introduce the ROI estimation framework followed by a case study. Finally, we conclude the chapter with a discussion and avenues for future research.

2.2 CLOUD ARCHITECTURE AND BUSINESS VALUE

Cloud computing adoption for business applications provides a number of potential benefits but the actual realisation of these benefits is not always straightforward. A careful evaluation of the suitability of different cloud solutions for a given business model or application is required. This is not a trivial task given the large number of cloud vendors and associated services available in the market. Despite the recent growth of different service models (Kächele et al. 2013), SaaS, PaaS and IaaS still account for the vast majority of the market (Gartner 2019). In this chapter, we specifically focus on IaaS and PaaS. These two service models, although different in nature, share a number of value drivers that users should explore when estimating the expected benefits of adoption. Figure 2.1 provides visual representation of the main differences between the traditional legacy technology stack and different cloud services.

The US National Institute of Standards and Technology (NIST) defines IaaS as:

> *The capability provided to the consumer is to provision processing, storage, networks, and other fundamental computing resources where the consumer is able to deploy and run arbitrary software, which can include operating systems and applications. The consumer does not manage or control the underlying cloud infrastructure but has control over operating systems, storage, and deployed applications; and possibly limited control of select networking components (e.g., host firewalls).* (Mell and Grance 2011, p. 3)

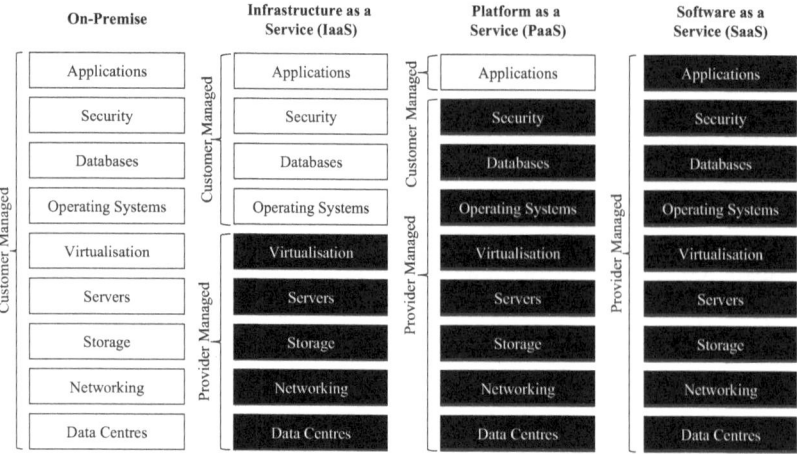

Fig. 2.1   Overview of different cloud services

As such, IaaS provides users with a high level of flexibility but requires a high level of IT skills in order to optimise and manage the infrastructure. In fact, developers are still required to design and code entire applications and IT administrators still need to install, manage, and integrate third-party solutions. Key benefits of IaaS are related to the fact that the typical tasks related to managing and maintaining a physical infrastructure are not required anymore, and additional infrastructure resources are available on demand and can be deployed in minutes instead of days or weeks (Kavis 2014).

PaaS is defined as:

> *The capability provided to the consumer is to deploy onto the cloud infrastructure consumer-created or acquired applications created using programming languages, libraries, services, and tools supported by the provider. The consumer does not manage or control the underlying cloud infrastructure including network, servers, operating systems, or storage, but has control over the deployed applications and possibly configuration settings for the application-hosting environment.* (Mell and Grance 2011, pp. 2–3)

PaaS sits on top of the cloud infrastructure and abstracts most of the standard application functions such as caching, database scaling, security, logging etc. and provides them as a service (Kavis 2014). Similar to IaaS, the user controls the self-installed applications but not the underlying infrastructure and platform. PaaS services mostly speak to developers as the PaaS vendors typically provide them with a suite of tools for speeding up the development process. Cloud platforms also facilitate the development of cloud native systems which, according to the Cloud Native Computing Foundation (CNCF 2018), are increasingly:

- Container-packaged;
- Dynamically managed by a central orchestrating process;
- Microservice-oriented.

Cloud native applications provide clear technical advantages in terms of isolation and reusability, which lower costs associated with maintenance and operations (Rosati et al. 2019). Both IaaS and PaaS are typically consumed by SaaS providers, which in turn offer their services to the final user in exchange for monthly or annual subscription fees (Cusumano 2008; Ojala 2012). In this context, a proper estimation of the TCO and ROI of the cloud represents the basis for adequate and effective pricing strategies, and for evaluating investment decisions (Rosati et al. 2019).

2.3   MEASURING THE ROI OF A CLOUD INFRASTRUCTURE

ROI is one of several financial metrics available to business decision makers to estimate the expected financial outcomes of an investment (Farbey and Finkelstein 2001; Rosati et al. 2017). While TCO focuses merely on costs, ROI includes both costs and benefits therefore providing a more forward-looking and comprehensive assessment of an investment. Despite the fact that the benefits of cloud computing extend well beyond cost savings, these have historically been the main drivers of adoption (CFO Research 2012). Unsurprisingly, TCO, rather than more strategic ROI, has been the main metric of cloud investment evaluation (Brinda and Heric 2017). TCO is attractive as, compared to ROI, it is easier to estimate, and cloud vendors make online TCO calculators available to their customers. However, these tools only focus on relatively simplistic tangible operational cost calculations (Rosati et al. 2017). A similar approach only provides a partial picture of the costs and benefits generated by cloud computing and may under- or over-estimate the financial outcomes of cloud investments and ultimately translate in to sub-optimal investments. To address this limitation, we present a more comprehensive framework for estimating the ROI of cloud investments for IaaS/PaaS (Fig. 2.2).

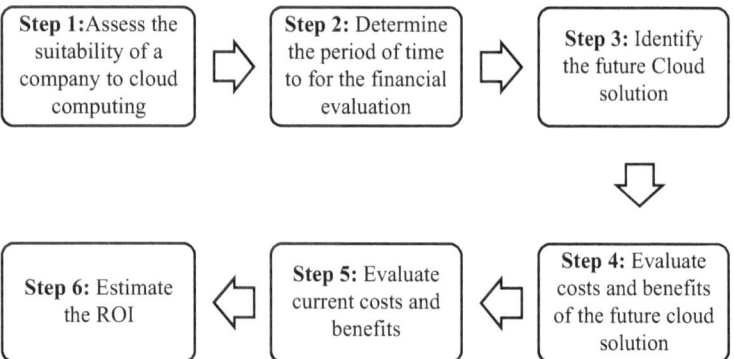

Fig. 2.2 Organisational ROI estimation framework for cloud computing investments

2.3.1 Step 1: Suitability to Cloud Computing

The initial phase of the ROI calculation is an assessment of the suitability of the organisation for the adoption of cloud computing. Despite the hype around cloud computing and the promising statements of cloud vendors, IaaS/PaaS adoption may not be the most effective and efficient solution for every organisation or every application deployed by or within an organisation (McKinsey and Company 2009; Misra and Mondall 2011). Misra and Mondall (2011) provide a weighted scoring model for estimating a suitability index. The model includes the following aspects:

- Size of IT resources and customer base characteristics: smaller organisations whose IT infrastructure is based in one country and that generate relatively limited amount of revenues from IT offering are more suitable to cloud computing than IT giants.
- The utilisation pattern of IT resources: cloud computing is particularly attractive for organisations with a highly variable workload profile as they can benefit from the on-demand scalability typical of cloud infrastructures.
- Sensitivity of the data handled by the organisation: cloud services may be riskier for organisations handling very sensitive data, particularly for applications running a public cloud.
- Workload criticality: highly critical workloads require more stringent, reliable and secure resources therefore it may be difficult to find a cloud vendor that is able to provide an adequate Service-Level Agreement (SLA).

The outcome of this initial step may prevent organisations that are clearly not suitable for the cloud from wasting additional resources in the evaluation process. The suitability index may also provide sort of a reality-check for estimated ROI as organisations that are more suitable for the cloud should expect a higher return on investment (Misra and Mondall 2011; Walterbusch et al. 2013).

2.3.2   Step 2: Determine the Period of Time for the Financial Evaluation

Five years is the typical time frame for estimating the ROI of large IT investments such as a cloud infrastructure. This is because the initial implementation requires time and resources; shorter time periods may not be long enough to capitalise such initial investment. Five years is not a fixed rule. Organisations should evaluate cloud investments within the most appropriate time frame for them considering the amount of investment, the overall expected duration of the investment, and its relationship with the overall strategic plan of the organisation. It should be said though that the longer the time frame the harder it becomes to estimate reliable figures associated with costs and benefits. This is particularly the case in fast-changing business environments where technologies, applications, and business models quickly become obsolete.

2.3.3   Step 3: Identify the Future Cloud Solution

The range of cloud computing offerings is very diverse and fragmented. This often makes very difficult to compare one cloud provider or service against others (Rehman et al. 2011). A number of different selection techniques and approaches have been developed over time which are more or less suitable for different cloud services (see Sun et al. (2014) for a detailed review). Regardless of the selection technique adopted, it is critical to identify a *to-be* solution that is directly comparable to the existing architecture. This does not mean that the two alternative architectures have to be comparable from a technical perspective. In fact, this may not be possible due to the different nature of cloud and on-premise solutions. However, they should be able to meet the same business requirements, and monetary values should be measured consistently across the two scenarios (ISACA 2012). Table 2.1 provides a list of key elements to consider during this phase.

Table 2.1   Key objectives of cloud service selection (adapted from ISACA (2012))

Objective	Guidance/key questions to answers
Define high-level business (functional) requirements	• What business functions need to be covered? • What are the business drivers for adopting cloud-based services? • How could cloud based services support business processes? • What compliance requirements are relevant?
Define a baseline cloud service model	• What type of cloud service model (e.g. IaaS, PaaS etc.)? • What kind of deployment model (e.g. public, private etc.)? • Where would the services be physically located? • Who would deliver the services? • Start with a model that is simple and low-cost and then exclude options that do not meet compliance and risk requirements.

Assess risks associated with the selected cloud model	• Identify risk areas to be considered (e.g. multitenancy, data usage limitations, security, privacy, migration costs etc.).
	• Determine countermeasures to mitigate the areas of risk outside the organisation's risk tolerance. These may include:
	  – Data encryption
	  – A revert-back strategy
	  – On-premise backups and audit trailing
	  – Clear and comprehensive SLA
	  – In-house disaster recovery strategy.

2.3.4   Step 4: Evaluate the Future Costs and Benefits

Costs and benefits evaluation is arguably the central activity of the ROI estimation. In this step, both operational and non-operational implications of cloud adoption should be taken into account. Costs can be grouped into three main categories i.e. upfront, recurring and termination costs. Table 2.2 provides an overview of the key cost components to be considered for each cost category. This list should not be considered as either rigid or exhaustive. Organisations should carry out their own assessment of potential direct and indirect costs associated with cloud adoption. For example, migration costs are not relevant for organisations aiming to design a greenfield cloud native application.

Table 2.2 Cost categories for the cloud

Category	Key cost components to consider
Upfront costs	• Start-up costs to prepare for the transition to the cloud • Technical/legal/consulting costs related to assessing/evaluating cloud alternatives and technical readiness • Network/bandwidth investments • Technical costs (including staff) for implementation/integration • Staff training • Change management
Recurring costs	• Cloud service(s) subscription fees • Cloud consumption costs (server, storage, database, network, throughput, CSP support fees) • Personnel costs (IT, finance, Human Resources)
Termination costs	• Costs relating to contract termination (legal/technical/consulting) • Early termination penalties • Alternative cloud service provider evaluation costs • Technical costs (data extraction/sanitisation) • Reinvestment or transfer back to on-premise (hardware acquisition and setup costs)

Similarly, the benefits generated by the adoption of cloud services can be grouped in to two main categories: tangible and intangible (ISACA 2012). Tangible benefits are clearly easier to identify and typically include additional revenues, faster time-to-market, lower operational costs etc. However, a significant portion of the value generated by the cloud adoption typically fall in to the second category. Figure 2.3 provides an overview of the potential benefits of cloud adoption for business applications. As per the cost drivers presented above, organisations should carry out their own assessment to identify which benefits may actually apply to their specific context.

2.3.5 Step 5: Evaluate the as-is Costs and Benefits

ROI estimation should be based on the comparison between two alternative scenarios. In the context of cloud adoption, the alternative scenario is typically an on-premise solution. Care needs to be taken when considering on-premise costs versus those in the cloud. While many are similar, there often subtle differences. Table 2.3 provides an overview of the key cost components to be considered for each cost category.

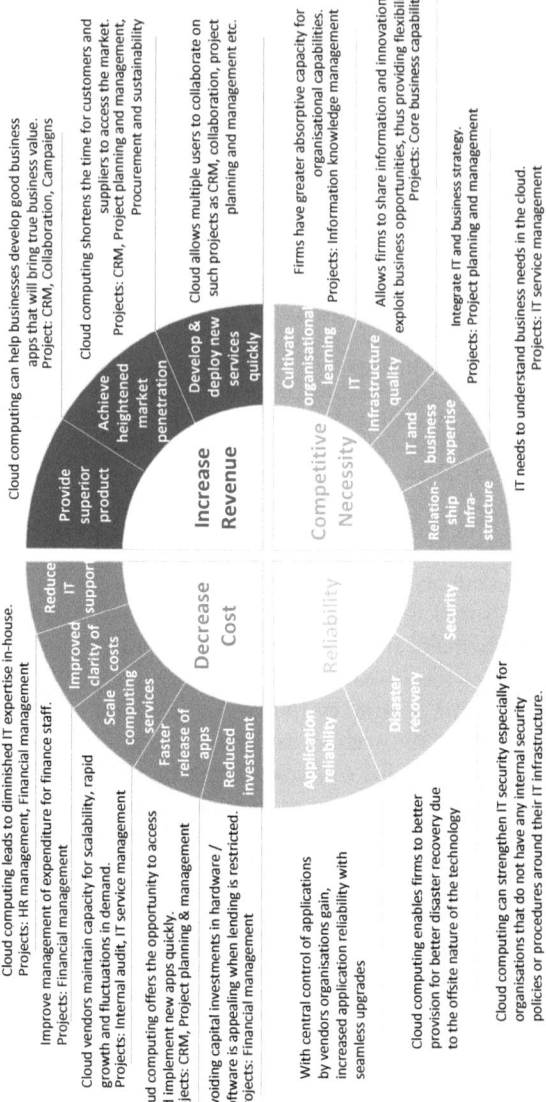

Cloud computing can help businesses develop good business apps that will bring true business value.
Project: CRM, Collaboration, Campaigns

Cloud computing shortens the time for customers and suppliers to access the market.
Projects: CRM, Project planning and management, Procurement and sustainability

Cloud allows multiple users to collaborate on such projects as CRM, collaboration, project planning and management etc.

Firms have greater absorptive capacity for organisational capabilities.
Projects: Information knowledge management

Allows firms to share information and innovation to exploit business opportunities, thus providing flexibility.
Projects: Core business capabilities

Integrate IT and business strategy.
Projects: Project planning and management

IT needs to understand business needs in the cloud.
Projects: IT service management

Cloud computing leads to diminished IT expertise in-house.
Projects: HR management, Financial management

Improve management of expenditure for finance staff.
Projects: Financial management

Cloud vendors maintain capacity for scalability, rapid growth and fluctuations in demand.
Projects: Internal audit, IT service management

Cloud computing offers the opportunity to access and implement new apps quickly.
Projects: CRM, Project planning & management

Avoiding capital investments in hardware / software is appealing when lending is restricted.
Projects: Financial management

With central control of applications by vendors organisations gain, increased application reliability with seamless upgrades

Cloud computing enables firms to better provision for better disaster recovery due to the offsite nature of the technology

Cloud computing can strengthen IT security especially for organisations that do not have any internal security policies or procedures around their IT infrastructure.

Fig. 2.3 IC4 Cloud computing strategic alignment model (Lynn 2018)

Table 2.3 Description and explanation of cost categories for on-premise

Category	Key components to consider

Upfront costs	• Large capital expenditure and investments in:
	  – Physical hardware and infrastructure
	  – Bandwidth
	  – Software
	  – Property and facilities, heating and cooling
	  – Staff training
	  – Procurements costs

Recurring costs	• Ongoing operational costs such as:
	  – Utilities (electricity, bandwidth)
	  – Premises and facilities (security, physical access, HVAC, electrical and UPS)
	  – Property costs (rent and rates)
	  – IT audits
	  – IT personnel costs (maintenance, admin, developer)
	  – Software/OS licenses

Termination (disposal) costs	• Disposal of physical hardware and infrastructure components • Depreciation • Compliance costs (secure data backup/cleansing) • Secure removal and disposal of IT and associated equipment

The benefits of the on-premise solution are then measured using the current performance of the organisation in terms of revenues, growth, customer satisfaction etc. Both costs and benefits of the on-premise solution represent the baseline for evaluating the incremental value or costs generated by the cloud adoption.

2.3.6   Step 6: Evaluate the as-is Costs and Benefits

The last step in the process consists of inserting all the numbers gathered in the previous steps in to an extended version of the ROI formula as presented in equation below:

$$
ROI = \frac{\left[\left(\text{Tangible Benefits} + \text{Intangible Benefits}\right) - \left(\text{Upfront Costs} + \text{Recurring Costs} + \text{Termination Costs}\right) \right]}{\left(\text{Upfront Costs} + \text{Recurring Costs} + \text{Termination Costs}\right)} \tag{2.1}
$$

where

$$\text{Cloud TCO} = \text{Upfront Costs} + \text{Recurring Costs} + \text{Termination Costs} \quad (2.2)$$

and

$$\text{Tangible Benefits} = \text{Incremental Revenues} + \text{Lower Costs} \quad (2.3)$$

In the case of a cloud migration rather than a greenfield cloud adoption, the additional value generated by the investment should be measured as the incremental value the cloud generates compared to the previous architecture. Therefore Eq. (2.1) would become:

$$\text{ROI} = \frac{\Delta \text{Gross Profit Margin}}{\text{TCO}} \quad (2.4)$$

where

$$\Delta \text{Gross Profit Margin} = \left(\text{Revenues}_{\text{Cloud}} - \text{TCO}_{\text{Cloud}} + \text{Additional Savings}_{\text{Cloud}}\right) \\ - \left(\text{Revenues}_{\text{Premise}} - \text{TCO}_{\text{Premise}}\right) \quad (2.5)$$

2.4   CASE STUDY

This section presents the application of this framework to a real-life study of a cloud infrastructure migration. We specifically focus on IaaS rather than PaaS adoption as the former requires estimating more cost components than the latter. As such, IaaS adoption provides a more comprehensive example that can then be adapted to a PaaS adoption scenario.

2.4.1   Company and Application Overview

The Company participating to this study operates in the financial services industry and provides a suite of applications for the delivery and support of financial and business technology solutions across EMEA, South America, Asia and Australasia. Through its development of proprietary technology, the Company has developed core products for currency conversion, multi-currency pricing, commercial and retail foreign exchange. In the year prior to the study, the Company reached almost €200 million in revenues and had almost 200 employees. The specific application being migrated has a customer and user base across Europe and the South Pacific region which includes both Business-to-Consumer (B2C) and Business-to-Business (B2B) customers.

When the forex exchange application was first developed, the Company did not consider cloud technology to be sufficiently sophisticated and capable of hosting it. They therefore decided to install it on an on-premise infrastructure in the Company's headquarters. The application was originally monolithic by design. However, over several years the Company incrementally migrated the application to a microservices architecture.

As the cloud evolved and adoption became mainstream, the Company has already moved its Continuous Integration and Quality Assurance environments into a containerisation model running on Microsoft Azure. These environments are managed by the Company's internal development team. In addition to the application servers, the cloud migration includes a container registry, a source code repository, a configuration server, and SQL databases. The development team also intend to use containerisation in its production environments. The Company indicated that the environments for user acceptance testing and production (currently managed by the internal infrastructure team) may also be migrated, depending of the outcome of the ROI estimation.

2.4.2   Suitability Index

The Company's main business drivers behind its decision to adopt the cloud are:

- Efficiency gains through automation of IT operations and implementation of site reliability engineering principles and practices;
- greater efficiency in development lifecycle;
- increased performance and reliability of applications;
- reduced cost;
- technical scalability to support business growth.

The Company's current IT setup is capable of supporting two million transactions per year across 1000 POS terminals while the Company requires the capability to scale to 30,000 terminals and 50 million transactions annually.

The initial step of the ROI calculation involves assessing the suitability of the Company to the cloud. A questionnaire was designed in order to capture all the information required to estimate the suitability index as proposed by Misra and Mondall (2011):

Size of the Company's IT resources: The infrastructure comprised a cluster of less than 100 servers and is hosted on-premise in a local data centre in the Company's headquarters. The customer base is geographically dispersed across Europe and the South Pacific and is served by the on-premise infrastructure. An indication of the size of Company's customer base can be derived from the scale of their operations, and their transaction volumes; the current system is capable of supporting 2 million transactions per year across 1000 POS terminals across the regions listed above.

Utilisation pattern of the resources: The Company may experience some peak surges by virtue of the fact that their system provides an online retail currency conversion service, and such transactions are typically seasonal in nature. As such, the Company's utilisation pattern could be profiled as having moderately variable workloads with occasional surges.

Sensitivity of the data they are handling: The Company classified the sensitivity levels of the data they handle as "sensitive" (personal information, contact details) and "very sensitive" (bank related data, transactional data). The data captured during customer transactions on the forex application is limited to the customer's name, address and proof of identification. The system does not handle and process online credit card payments thus credit card information is not stored. As such, the company has no PCI DSS compliance requirements. Currently, payments are processed using the 3D secure authentication standard with a third-party service provider. There are other related compliance requirements for service management and customer value (ISO 20000-1) and information security management (ISO 27001) that the Company has to adhere too.

Workload criticality: The Company indicated that migrating the RFX application was highly critical. A primary reason for this is the ease with which cloud services enable firms to easily and efficiently handle potential failover situations, thereby preserving business continuity and preventing data loss. This circumvents and eliminates the internal administrative overhead required for presenting the business case for the purchase of additional hardware.

Given the information provided above and the weights proposed in Misra and Mondall (2011), the Company obtained a suitability index of 3876 which falls within the intermediate category (Misra and Mondall 2011). This suggests that further investigation such as an ROI study is required before deciding to adopt the cloud infrastructure.

2.4.3   ROI Estimation

A five-year time frame was adopted for the ROI study and the Company was asked to fill in a detailed questionnaire in order to identify the required costs and expected revenues associated with both the current on-premise solution and the alternative cloud infrastructure. This phase of the study spanned over three months and required the involvement of ten people across five different departments i.e. top management, IT, finance, business unit, and human resources.

Both the cloud (*to-be*) and the on-premise infrastructure (*as-is*) were designed to deliver the same amount of revenues over the time period of the analysis. Therefore, any change in value has to be driven by the cost reduction and/or intangible benefits. Table 2.4 summarises the TCO calculation for both scenarios.

The cloud infrastructure is expected to generate a cost saving of €352,464 over five years mostly due to the lower upfront costs and no termination costs. In fact, the estimated upfront costs of the cloud solution only included IT training (€9000) and cloud assessment and consulting costs (€20,000). The company also identified a number of potential intangible benefits related to cloud migration such as:

- enhanced productivity;
- improved compliance and security;
- the ability to focus on core business;
- access to the cloud provider's expertise and capabilities.

These suggest an approximate total net positive cost saving of €81,000. This ultimately results in an expected ROI of:

$$\text{ROI} = \frac{€352,464 + €81,000}{€3,086,188} = \frac{€433,464}{€3,086,188} = 14.05\%$$

Table 2.4   TCO summary

On-premise		Cloud	
Upfront on-premise cost	€114,500	Upfront cloud cost	€29,000
Recurring on-premise cost	€3,316,652	Recurring cloud cost	€3,057,188
Termination on-premise cost	€7500	Termination cloud cost	€0
TCO on-premise	€3,438,652	TCO cloud	€3,086,188

Based on the positive, although not very high, expected ROI and considering other further strategic considerations, the Company decided to migrate the RFX application to the cloud. The Company also realised that an unforeseen benefit arising from the project was the increased transparency and comparison of the costs associated with their on-premise infrastructure and that of their cloud service consumption. This may help with decision making and developing the business case for future considerations when deciding between a reinvestment in on-premise hardware or adopting cloud services.

2.5   CONCLUSION

In this chapter, we presented a practical framework for estimating the return on cloud computing investments from the customer perspective. We focused specifically on Infrastructure-as-a-Service and Platform-as-a-Service as they have more extensive organisational implications than simpler SaaS applications. Several online tools and methodologies based on relatively simplistic tangible operational cost calculations have been made available to firms for calculating the total cost of ownership of their cloud investments. However, cloud services also generate intangible costs and benefits that should be taken into account when embarking on the cloud journey. Investment decisions taken on the basis of partial assessments of the potential business value generated by cloud adoption may result in sub-optimal budget and capital allocation and ultimately undermine an organisation's competitive advantage. Our framework aims to overcome such limitations by providing a step-by-step process for estimating a comprehensive ROI of cloud adoption. The actual implementation of the framework is shown though a real-life case study of infrastructure migration.

Future research may explore how the ROI estimation framework presented above can be adapted to different cloud migration scenarios (Jamshidi et al. 2013) or to relatively new paradigms of cloud computing such as serverless computing (or Function-as-a-Service—FaaS) (Lynn et al. 2017). Finally, further studies may also investigate the relationship between the adoption of more comprehensive ROI measures to and the effectiveness of IT investment decision-making.

REFERENCES

Armbrust, Michael, Fox Armando, Rean Griffith, D. Joseph Anthony, K. Randy, K. Andy, Gunho Lee, et al. 2010. A View of Cloud Computing. *Communications of the ACM* 53 (4): 50–58.

Brender, Nathalie, and Iliya Markov. 2013. Risk Perception and Risk Management in Cloud Computing: Results from a Case Study of Swiss Companies. *International Journal of Information Management* 33 (5): 726–733.

Brinda, Mark, and Michael Heric. 2017. The Changing Faces of the Cloud. https://www.bain.com/insights/the-changing-faces-of-the-cloud.

Cegielski, Casey G., L. Allison Jones-Farmer, Yun Wu, and Benjamin T. Hazen. 2012. Adoption of Cloud Computing Technologies in Supply Chains: An Organizational Information Processing Theory Approach. *The International Journal of Logistics Management* 23, no. 2: 184–211.

CFO Research. 2012. The Business Value of Cloud Computing—A Survey of Senior Finance Executives. http://lp.google-mkto.com/rs/google/images/CFO%2520Research-Google_research%2520report_061512.pdf.

Cloud Native Computing Foundation (CNCF). 2018. Cloud Native Definition v1.0. https://github.com/cncf/toc/blob/master/DEFINITION.md.

Cusumano, Michael A. 2008. The Changing Software Business: Moving from Products to Services. *Computer* 41 (1): 20–27.

Farbey, Barbara, and Anthony Finkelstein. 2001. *Evaluation in Software Engineering: ROI, but More Than ROI*. 3rd International Workshop on Economics-Driven Software Engineering Research (EDSER-3), ON, Canada.

Farbey, Barbara, Frank William Land, and David Targett. 1993. *How to Assess Your IT Investment: A Study of Methods and Practice*. Boston: Butterworth-Heinemann.

Gartner, Inc. 2019. Gartner Forecasts Worldwide Public Cloud Revenue to Grow 17.5 Percent in 2019. Accessed 2 April 2019. https://www.gartner.com/en/newsroom/press-releases/2019-04-02-gartner-forecasts-worldwide-public-lic-cloud-revenue-to-g.

IDG Communications. 2018. Cloud Computing Survey. Accessed 14 August 2018. https://www.idg.com/tools-for-marketers/2018-cloud-computing-survey/.

ISACA. 2012. Calculating Cloud ROI: From the Customer Perspective. http://www.isaca.org/Knowledge-Center/Research/ResearchDeliverables/Pages/Calculating-Cloud-ROI-From-the-Customer-Perspective.aspx.

Jamshidi, Pooyan, Aakash Ahmad, and Claus Pahl. 2013. Cloud Migration Research: A Systematic Review. *IEEE Transactions on Cloud Computing* 1 (2): 142–157.

Kächele, Steffen, Christian Spann, Franz J. Hauck, and Jörg Domaschka. 2013. *Beyond IaaS and PaaS: An Extended Cloud Taxonomy for Computation, Storage and Networking*. 2013 IEEE/ACM 6th International Conference on Utility and Cloud Computing, 75–82. IEEE.

Kavis, Michael J. 2014. *Architecting the Cloud: Design Decisions for Cloud Computing Service Models (SaaS, PaaS, and IaaS)*. John Wiley & Sons.

Leimbach, Timo, Hallinan Dara, Bachlechner Daniel, Weber Arnd, Jaglo Maggie, Øjvind Nielsen Rasmus, Nentwich Michael, Strauß Stefan, Hunt Graham, and Lynn Theo. 2014. Potential and Impacts of Cloud Computing Services and Social Network Websites. https://www.europarl.europa.eu/RegData/etudes/etudes/join/2014/513546/IPOL-JOIN_ET(2014)513546_EN.pdf.

Lynn, Theo. 2018. Addressing the Complexity of HPC in the Cloud: Emergence, Self-Organisation, Self-Management, and the Separation of Concerns. In *Heterogeneity, High Performance Computing, Self-Organization and the Cloud*, 1–30. Cham: Palgrave Macmillan.

Lynn, Theo, Pierangelo Rosati, Arnaud Lejeune, and Vincent Emeakaroha. 2017. *A Preliminary Review of Enterprise Serverless Cloud Computing (Function-as-a-Service) Platforms*. 2017 IEEE International Conference on Cloud Computing Technology and Science (CloudCom), 162–169. IEEE.

Marston, Sean, Zhi Li, Subhajyoti Bandyopadhyay, Juheng Zhang, and Anand Ghalsasi. 2011. Cloud Computing—The Business Perspective. *Decision Support Systems* 51 (1): 176–189.

McKinsey and Company. 2009. Risultati di ricerca Risultati Web Clearing the Air on Cloud Computing. https://www.dpcinc.com/pdf/ClearingtheAironthe Clouds.pdf.

Mell, Peter, and Tim Grance. 2011. The NIST Definition of Cloud Computing. https://nvlpubs.nist.gov/nistpubs/Legacy/SP/nistspecialpublication800-145.pdf.

Misra, S. C., & Mondal, A. (2011). Identification of a Company's Suitability for the Adoption of Cloud Computing and Modelling its Corresponding Return on Investment. *Mathematical and Computer Modelling*, 53 (3–4): 504–521.

Ojala, Arto. 2012. *Software Renting in the Era of Cloud Computing*. 2012 IEEE Fifth International Conference on Cloud Computing, 662–669. IEEE.

Rehman, ur Zia, Farookh K. Hussain, and Omar K. Hussain. 2011. *Towards Multi-criteria Cloud Service Selection*. 2011 Fifth International Conference on Innovative Mobile and Internet Services in Ubiquitous Computing, 44–48. IEEE.

Ronchi, Stefano, Alessandro Brun, Ruggero Golini, and Xixi Fan. 2010. What is the Value of an IT e-Procurement System? *Journal of Purchasing and Supply management* 16 (2): 131–140.

Rosati, Pierangelo, Grace Fox, David Kenny, and Theo Lynn. 2017. *Quantifying the Financial Value of Cloud Investments: A Systematic Literature Review*. 2017 IEEE International Conference on Cloud Computing Technology and Science (CloudCom), 194–201. IEEE.

Rosati, Pierangelo, Frank Fowley, Claus Pahl, Davide Taibi, and Theo Lynn. 2019. *Right Scaling for Right Pricing: A Case Study on Total Cost of Ownership*

Measurement for Cloud Migration. Cloud Computing and Services Science: 8th International Conference, CLOSER 2018, Funchal, Madeira, Portugal, March 19–21, 2018, Revised Selected Papers, vol. 1073, p. 190. Springer.

Strebel, Jörg, and Alexander Stage. 2010. *An Economic Decision Model for Business Software Application Deployment on Hybrid Cloud Environments.* Multikonferenz wirtschaftsinformatik, Göttingen, Germany, vol. 2010, p. 47.

Sun, Le, Hai Dong, Farookh Khadeer Hussain, Omar Khadeer Hussain, and Elizabeth Chang. 2014. Cloud Service Selection: State-of-the-Art and Future Research Directions. *Journal of Network and Computer Applications* 45: 134–150.

Walterbusch, M., B. Martens, and F. Teuteberg. 2013. Evaluating Cloud Computing Services from a Total Cost of Ownership Perspective. *Management Research Review* 36 (6): 613–638.

CHAPTER 3

The SaaS Payoff: Measuring the Business Value of Provisioning Software-as-a-Service Technologies

Trevor Clohessy, Thomas Acton, and Lorraine Morgan

Abstract Creating and capturing value with new digital technologies such as cloud computing is often fraught with complexity and ambiguity for incumbent information technology (IT) firms. Using the business model concept as a lens, the objective of this chapter is to address a current gap in our knowledge about the impact of Software-as-a-Service (SaaS) on incumbent IT supply-side organisations. The empirical findings from a cross-case study analysis of two incumbent IT service providers lead to a number of in-depth insights that are discussed in this paper. The study

T. Clohessy (✉)
Department of Enterprise and Technology, Galway-Mayo Institute of Technology, Galway, Ireland
e-mail: trevor.clohessy@gmit.ie

T. Acton • L. Morgan
Business Information Systems Department, J.E. Cairnes School of Business & Economics, National University of Ireland, Galway, Ireland

39
T. Lynn et al. (eds.), *Measuring the Business Value of Cloud Computing*, Palgrave Studies in Digital Business & Enabling Technologies, https://doi.org/10.1007/978-3-030-43198-3_3

identifies six tangible business model payoffs that have resulted from provisioning SaaS technologies. Subsequently, this paper lays the foundation for contributing to understanding how SaaS technologies can influence business models.

Keywords Cloud computing • Software-as-a-Service (SaaS) • Business model • Payoffs

3.1   Introduction

It is treacherous on a tightrope to change your focus point and suddenly look down. Philip Pettit (The Walk 2015)

This above quote rings no truer than in the current digital technological arena which is characterised by rapid fluctuation and turbulence—a fluid landscape where a multitude of incumbent information technology (IT) organisations are having to change their digital focus and "look down" and embrace emerging digital technological advancements. Digital transformation is concerned with the changes digital technologies can bring about in an organisation's business model and the subsequent changes in products, organisational structures and the automation of processes (Clohessy et al. 2017; Benlian et al. 2016; Hess et al. 2016). The business model concept has been used extensively to examine how digital technologies transform organisational abilities to create and capture value (e.g. the internet, e-commerce platforms, mobile applications, Big Data, analytics, and so on). Driving factors such as the emerging knowledge economy, the restructuring of global financial services, increased outsourcing of business processes and information systems, rapid advancements in digital technologies and the repeated failure of organisations to capitalise on the capabilities afforded by these technologies have catapulted the business model concept back into the public arena (Peters et al. 2015).

In the past decade corporate investments in cloud computing, specifically Software-as-a-Service (SaaS), has increased and become a substantial component of business. Examples of personal use SaaS include, for example, Gmail, Skype, and Dropbox. Salesforce customer relationship management (CRM), SAP Analytics Cloud, and Oracle Enterprise Resource Planning Cloud are examples of business-centric SaaS. Often mentioned benefits include the reduced need for up-front investments in IT infrastructure, IT software and IT skills, reduced costs and enhanced flexibility (Guo and Ma 2018). SaaS accounts for the majority of the entire cloud market and it is forecast to reach $117.1 billion in revenues by 2021 with an annual growth rate of 16 percent (Gartner 2018). Using the business model as an anchor, this chapter provides new insights into the business model payoffs incumbent IT service providers have been able to leverage as a result provisioning SaaS technologies. Such insights can pave the path for establishing significant theoretical contributions for understanding underlying mechanisms of business model success with regards to SaaS. Our study was guided by the following question:

What business model payoffs manifest for incumbent IT service providers because of provisioning Software-as-a Service technologies?

The remainder of the chapter is structured as follows. In Sect. 3.2, we first provide a rationale for the study and provide an overview of the business model concept. Section 3.3 describes the case study methodology used to address the question above. Section 3.4 discusses the results of the case studies analysed. Finally, the paper concludes with an outline of the study's limitations and the broader study implications where we outline how this research extends extant theoretical and practical contributions.

3.2   Study Background

3.2.1   SaaS Provision

There has been an emergent body of research focused on investigating the impact of SaaS on organisational business models. While the majority of this research has been carried out from an adoption perspective, there is a dearth of research which has also explored the impact from an IT supply-side perspective. Thus, the manner with which these organisations are attempting to unleash the digital transformative potential of SaaS through their business models is an area which merits further scrutiny (Benlian et al. 2016; Hess et al. 2016). For instance, there is anecdotal evidence to suggest that incumbent IT service providers are experiencing substantial difficulties in their endeavours to leverage the business model payoffs of provisioning SaaS. This is evidenced by IT stalwarts such as Dell, Intel, IBM and Hewlett Packard (HP) whose struggles pertaining to how to best leverage the payoffs have been well documented.

3.2.2   Bounding the Business Model Concept

It has been argued that the utilisation of the business model concept as an anchor for the identification of the impact of new technologies on organisations is a fairly novel endeavour; it also remains an area which is under researched (Díaz-Díaz et al. 2017). In light of the comprehensive digitisation of enterprises at large, this seems all the more surprising. Business models not only serve as instruments for digital strategic planning but are also used for developing new and existing business activities (Van Kerkhoff et al. 2014). While the single components of extant business model frameworks vary extensively in the literature (Wirtz et al. 2016), they do converge to four overarching dimensions (e.g. value proposition, value co-creation, value delivery and value capture) which can be used to analyse, describe and classify the constituent parts of business models (Peters et al. 2015). In order to address our research questions, we used the Service, Technology, Organisation and Finance (STOF) business model framework (Bouwman et al. 2008) which typifies these four overarching dimensions (Table 3.1). The STOF framework describes how a network of cooperating organisations create and capture value from new digital services across four core business model domains (service, technological, organisational and financial).

The service domain is directly related to the value that is derived by the provider and customer from the service offering. The service offering must be considered better and deliver the desired satisfaction more effectively and efficiently than competitors; customer or user experience is key

Table 3.1   STOF business model research framework (Bouwman et al. 2008)

Business model domain	Description
Service domain	Delineates an organisation's service offering and the inherent value propositions and the specific end-users in particular target customer segments.
Technological domain	Describes the technical functions and core competencies needed to realise the service offering.
Organisational domain	Defines how the organisation creates value from a service offering via the configuration of actors (value network) comprising resources which together perform value activities.
Financial domain	Conveys the revenue and cost structure arrangements operationalised in order to capture value from a service offering.

(Bouwman et al. 2008). Functionality and technical architecture play pivotal roles in the technology domain. Functionality refers to the range of operations that can be performed by the service offering. The technical architecture relates to the "overall architecture of the components of a technical system in terms of backbone infrastructure, devices, service platforms, access networks and applications" (Bouwman et al. 2008, p. 115). The organisational domain revolves around the concept of a value network comprised of actors who possess "certain resources and capabilities, which interact and together perform value activities, to create value for customers and to realise their own strategies and goals" (Bouwman et al. 2008, p. 116). Finally, the financial domain delineates how value is captured by various actors in a value network. This domain focuses on financial arrangements which *"revolve around investment decisions, revenue models, and revenue sharing arrangements...[and] are aimed at average cost-effectiveness, net cash worth, and internal return"* (Bouwman et al. 2008, p. 116).

The STOF framework is useful for a variety of reasons. First, it is relatively comprehensive, coherent and comprises business model components which are similar to other widely cited categorisations such as the business model canvas (Osterwalder and Pigneur 2010), the V^4 business model ontology (Al-Debei and Fitzgerald 2010) and the integrated business model (Wirtz 2011). Second, the STOF framework typifies the business model elements contained within the concept matrix proposed recently by Peters et al. (2015). Third, it has been previously utilised to assess the impact of SaaS technologies on business models (Lee et al. 2014) although, it should be noted that these aforementioned studies have not focused on incumbent IT service providers. Finally, the STOF framework is dynamic in nature as it encapsulates external factors of influence in terms of market dynamics, technological advancements and regulatory changes that all represent salient factors in the context of provisioning SaaS technologies.

3.3   Methodology

A multi-method, comparative case study research design was selected for the study. Table 3.2 provides an overview of the primary data sources (i.e. respondent interviews) and the secondary data sources that were analysed as part of the case study. The two case firms selected for the study provide rich environments for investigating our research objective. They are large

Table 3.2   Study data sources

Primary data sources (20 interviews)

ID	Industry/size/business model	Role	Industry experience (years)
A	*Software/Large ITSP/Mature*		
CA1		Senior SaaS Architect	8
CA2		SaaS Strategy Leader	9
CA3		SaaS Product Manager	12
CA4		Senior Cloud Infrastructure Developer	15
CA5		SaaS Leader	11
CA6		SaaS Strategy Leader	6
CA7		Chief Technology Officer	9
CA8		SaaS Product Manager	9
CA9		SaaS EMEA Leader	11
B	*Software/large ITSP/mature*		
CB1		Research & Development Director	20
CB2		Senior SaaS Architect	7
CB3		Senior SaaS Engineer	19
CB4		EMEA SaaS Leader	13
CB5		Cloud Datacenter Manager	6
CB6		Senior SaaS Manager	17
CB7		Senior SaaS Technologist	14
CB8		Senior SaaS Engineer	11
CB9		Senior SaaS Manager	9
CB10		SaaS Development Manager	12
CB11		SaaS Product Manager	18

Secondary data sources

Websites, white papers and marketing materials	Annual and quarterly reports	Company presentations, Blogs, YouTube, Webinars and Podcasts	Industry commentary and analysis and newspaper articles	Researcher's field notes	Reflective memos

(>10,000 employees) multi-national incumbent IT service providers who have been at the forefront of the advancement and provision of SaaS technologies for the past six years. As such, both cases represent theoretical sampling (Glaser and Strauss 1967) and make it suitable for analytical generalisation (Yin 2003). Furthermore, developing insightful narratives for the digital age, calls, in part, for the selection of *"compelling cases...[which are capable of producing]...powerful intellectual accounts"* (Henfridsson 2014, p. 1). For company confidentiality, we pseudonymously refer to both case study companies as Case A and Case B. Following the standard practice of using senior management as data sources (Klein and Myers 1999; Flyvbjerg 2006; Iyer and Henderson 2012), we selected senior managers from each case organisation. A case study approach to analyse emergent complex field problems more than anything else, requires experience (Flyvbjerg 2006). As such, the interviewees were selected based on the following criteria: first, the respondents should have experience working with SaaS technology. Second, the respondents should hold managerial positions (e.g. SaaS product manager, chief technology officer, etc.), which would enable them to have an in-depth knowledge of the business model intricacies of their SaaS operations. Third, the respondents should preferably have responsibility for overseeing their organisation's business model activities. Table 3.2 provides an overview of the 20 respondent's cloud roles and their number of years' IT industry experience.

3.4   Discussion of Findings

Prior to provisioning SaaS technologies, both Case A and Case B core business activities encompassed the manufacturing and distribution of enterprise servers, storage devices and a diverse range of computational software. These companies have an illustrious heritage pertaining to their ability to innovate their business models in order to leverage nascent technological advancements. Since 2015, both companies have prioritised the realignment and restructuring of their traditional IT business activities to focus solely on the provisioning of best of breed SaaS IT services. Therefore, both case organisations have experienced substantial success in the market. Table 3.3 provides a summary of the six tangible payoffs identified from the data analysis along the core business model domains. These payoffs can be categorised as being economic, business and transformative.

In terms of the *service business model domain* payoffs, the findings reveal that SaaS facilitates the provision of new products and services and enables an extended market reach. Concerning the former transformative benefit, the analysis revealed that SaaS has facilitated (1) the provision of virtualised SaaS-based solutions from a centralised location which possess the capability to automatically scale-up and down dynamically based on the

Table 3.3   Tangible business model payoffs from provisioning SaaS

Business model payoff	Empirical evidence from our study
Service domain: Provision of New Products and Services (Transformative)	SaaS enables both case organisations to create new and innovative business ventures beyond their existing traditional hardware and software services. The five essential characteristics (e.g. on demand self-service, measured service etc.) which underpin the SaaS model enables the case organisations to deliver nuanced and customisable value propositions to the customer.
Service domain: Extended Market Reach (Business)	SaaS has enabled both organisations to penetrate new horizontal and vertical market segments. The case organisations have established new growth strategies (e.g. new SaaS leader's roles, SaaS marketing teams, digital ecosystems) within these new segments in order to further establish their presence.
Technological Domain: Fast Software Development, Deployment and Maintenance (Transformative)	Centralised data centres and advancements in automation and scalability have transformed the manner with which both case organisations develop, deploy and maintain IT services. This transformation has enabled the case organisations to change the polarity of how they conduct business with customers.
Organisational Domain: Enhanced Agility (Business)	Both case organisations are pivoting rapidly towards agile methodologies in order to effectively create value from provisioning SaaS technologies. Both companies have experienced enhanced business agility as a result of their large scale internal restructuring of their existing departments, teams, developmental practices and collaborative tools in line with their SaaS developmental strategies.
Organisational Domain: Expanded Value Network (Transformative)	The results revealed that in order to step in line with the orientation of the SaaS market towards hybrid, open, and interoperable SaaS services both case organisations have had to carry out substantial restructuring of their traditional static and rigid value networks. These new SaaS value networks comprise a multitude of new actors and practices.
Financial Domain: Reduced Operating Costs (Economic)	SaaS has enabled the case organisations to significantly reduce their operating costs in comparison to the traditional mode of operation. The ability to centralise their provisioning operations from key global locations was identified as a key contributor to this reduced cost. The subsequent savings are being reinvested by both organisations into new technological strategic priorities.

demand for computational resources and (2) the creation of new and innovative products which have subsequently led to new spin-off services. The advantages derived from these new products and services (e.g. new revenue streams, reduced development times frames and budgets) would not have been feasible without SaaS. For example, senior managers from Case A discussed how SaaS has enabled the company to create a suite of new products and services. For example, they all pointed to one of their most successful products which is currently receiving global recognition for its cognitive capacity. The new system, which encompasses state of the art real time big data analytics functionality, is currently being used extensively in medical, pharmaceutical and biotechnology industry sectors. A senior manager described how the product was also being used at global sporting events:

> We are using the SaaS system to help us monitor major sporting events in terms of internet traffic, social analytics, and sentiment analysis. Based on these metrics the cognitive system can compute whether or not these factors will cause a spike in the usage of the customer services (e.g. website, booking systems etc.). The system then automates additional headroom on the capacity to cater for this spike without any human intervention. (CA5)

The informants confirmed that without SaaS technology, this product's core value propositions would be not as attractive for the customer. As one senior manager remarked:

> SaaS has enabled us to create a virtualised product which is 80% smaller, roughly 20 times faster and possesses exponentially more functionality in comparison to if we had attempted to design it with traditional methods. This product has been the catalyst for new spin-off cloud and Big Data services. (CA9)

In terms of the *extended market reach* benefit, the study revealed that SaaS enabled both organisations to enhance services to existing market segments while concurrently penetrating both new horizontal and vertical market segments. The case informants described how high costs, long project implementation periods and rigid partner networks encompassed within their traditional business models represented salient barriers to expanding their market reach. However, the findings identified that SaaS all but eradicated these barriers and enabled both organisations to not only provision SaaS services directly from their indigenous website and digital ecosystem portals but also configure new virtual value networks in order to access new customers across diverse industry segments. Individual customers and small and medium enterprises represent new market segments which both case organisations are attempting to establish a strong presence. In order to effectively leverage the extended market reach business benefit, both companies have established new growth strategies (e.g. new SaaS leader roles, SaaS marketing teams, digital ecosystems) within these new market segments.

Senior Managers from Case A and Case B provided the following insights:

> *SaaS technology has enabled us to provide our offerings to a broader market in comparison to our traditional mode of operation. Traditionally, numerous business partners along the value network would supply and install our products. We can now provision these same offerings rapidly from a centralised location which has dramatically reduced our costs. (CB11)*

> *SaaS has opened up new markets in terms of acquiring new customers who are interested in solely SaaS based solution and also providing SaaS services to our existing customer base. With our particular SaaS products, we are able to distribute it to multiple customers in a multi-tenant environment. Should our customer base grow we can automatically provision new servers and create working environments for them within an hour. (CA4)*

In terms of the transformative technological business model domain benefit of *fast software development, deployment and maintenance*, this study identified how centralised data centres and advancements in automation and scalability have transformed the manner with which both case organisations develop, deploy and maintain IT services. This transformation enabled the case organisations to change the polarity of how they conduct business with customers. SaaS enables IT service providers to consolidate multiple customers into a single centralised data centre location that is managed by a core group of employees. SaaS also significantly transformed both case organisation's automation capabilities by enabling them to roll out products, services and features rapidly. This capability is extremely important given the variances in customer requirements which might require frequent changes to SaaS solutions. The case organisations can now upgrade their SaaS offerings to their latest versions seamlessly from a centralised location. In some instances, this process can occur without any human intervention. In the traditional model, the case organisation's IT products/infrastructure would have been dispersed globally which would have incurred substantial time and cost constraints. The following insights by senior managers from Case A and Case B sums up nicely the service transformation that is taking place in both organisations from a technological domain point of view:

> *It is not traditional IT any more. It is all about fast repurposing. In the traditional IT model, if a machine went down, it could take a couple of days or even weeks to repair. Now in a SaaS computing context if a virtual machine goes down another one can be quickly spun up in its place. You can just nuke the old machine and create a new one. There is flexibility and scalability from a provider's point of view to accommodate the multifarious needs of our customers in real time 24/7/365. Our CPU cycles are being repurposed constantly. (CB7)*

> *In our traditional model, we were limited by the tight time frames with which we would have had to adhere to in order to develop and deploy a particular product. There would have been absolutely no lee way given what so ever. The final product delivered which was delivered was oftentimes substandard which would subsequently have led to a lot of negative press. If we had used SaaS for similar projects, we would have been able to develop and release the products far more rapidly leading to a greater product success. (CA3)*

Enhanced agility and an *expanded value network* were identified as the two keys organisational business model domain payoffs derived from operationalising SaaS-enabled business models. With regards to the former business benefit, both case organisations are currently undergoing large scale internal and external SaaS transformation. Their long-term objective is to provide the majority of their portfolios of capabilities in SaaS service models formats (e.g. high-end consulting, technical services, business processes, software and so on). The companies are also carrying out an internal restructuring of all of their existing teams, developmental practices and collaborative tools in line with their SaaS developmental strategies. This transformation not only resulted in cost savings but also enabled both case organisations to enhance their business agility. The findings revealed that it was a deliberate strategy for both organisations to address the serious agility gaps which were curtailing their abilities to respond effectively to a rapidly changing technological landscape. At that time both organisations existing levels of business agility were ineffectual at coping with the nuances inherent to provisioning SaaS technologies. Thus, strategic developments were prioritised and set in motion in order to address these agility gaps. This enhanced agility not only enabled both case organisations to deal effectively with a continually evolving and increasingly uncertain technological landscape, but also resulted in improved internal collaboration practices within both organisations. Senior managers from Case A and Case B provided the following insights:

> *The company as a whole have readily embraced the SaaS movement. In the last five years, SaaS has become pervasive across all of our business units. Traditionally we were seen as being the equivalent of a large cargo boat of the IT world. It wasn't sexy but we got the job done. We were a safe choice. However, with SaaS the customer does not want the large freight, they want us to be a racing yacht encompassing the same level of robustness, but they want that service for the price of a renting a row boat. That is why everything we do has to be SaaS native. (CA8)*

> *SaaS has completely changed the paradigm of how we do business whereby it has significantly enhanced our agility. We are not only responding to customer needs and suppliers faster but are also able to react to competitors more effectively. We are also using our agile war experiences in order to help our customers maximise SaaS enabled agility within their organisations. (CB6)*

With regards to the transformative *expanded value network* benefit, these value networks are paramount for creating value with SaaS. For example, SaaS facilitated both case organisations to create new flexible virtual value networks in order to shape attractive value propositions and revenue streams which would not be feasible for both organisations on their own. As a senior manager from Case A remarked:

> *SaaS has encouraged us to reimagine what our traditional value network, which was largely rigid and closed off to a small number of business partners, would look like on a virtual plain which is open, ubiquitous, flexible and contains thousands of actors. (CA3)*

Both case organisations have recently developed indigenous OpenStack collaborating network platforms which enables service providers, independent developers, resellers, integrators and telecommunications companies to resell both case organisation's SaaS products and services. These newly formed partnering programs enable them to enhance their ability to target customers on a global scale who favour open source, interoperable and hardware agnostic SaaS services. The informants also reported that customers are increasingly playing a more prominent role within their SaaS value networks. This is enabling both case organisations to create best of breed SaaS products and solutions and bring them to market faster. Both case organisations provide customers with open tools and services in order to transform their standardised products to align with the multifarious nature of customer requirements. The participants also revealed that their organisations hold regular networking events at which customers can give them face-to-face feedback on their services. Both case organisations have also simplified the process with which business partners along the value network can demo their SaaS solutions to customers.

The final SaaS-enabled business model benefit relates to the financial domain whereby provisioning SaaS technologies facilitated *reduced operating costs*. This economic benefit manifested in the significant reduction of their operating costs in comparison to their traditional mode of operation. SaaS enabled the centralisation of provisioning operations from key global locations (e.g. management of SaaS services in terms of monitoring and providing service upgrades). This is in comparison to the traditional mode of operation which encompassed the running of small silos of compute across both organisations, which were maintained by different teams. This ability to centralise costs while simultaneously achieving an increase in customer volume was identified as a key contributor to this reduced cost. The analysis also revealed that the costs pertaining to customer relationship management were also significantly impacted. First, SaaS significantly reduced both firm's costs pertaining to acquiring new customers. The flexibility inherent to the SaaS-enabled economic models, whereby customers can now trial or purchase SaaS solutions via credit card, purchasing order, or finance methods, has been revolutionary for both companies. The whole process is seamless in comparison to the traditional model. Second, provisioning SaaS technology significantly lowered both companies' administrative overheads pertaining to how they manage and support customers. Both organisations provide their customers with the ability to independently configure and manage SaaS services (e.g. OpenStack). Self-support facilities are also provided to enhance the simplicity and ease of use.

Senior managers from Case A and Case B also commented on the economies of scale which were being derived from provisioning centralised SaaS services:

From the company's point of view, it is far more cost effective for us to operationalise one data centre which manages a thousand customers rather than having to partner on a thousand different solutions and assigning those with individual organisational staff. So, collapsing a totality of needs into one centralised data centre manifests in cost efficiencies in service development, management, distribution and resourcing. (CA7)

Over time we envisage more and more compute resources being concentrated in fewer but larger data centres. The cost of making hardware and servers, in the traditional model, which could be used and serviced safely by a regular user was quite high. Now these are being aggregated in large scale data centres. The company are optimising for very large data centres where we have complete control from a centralised location. (CB1)

The case study analysis also revealed that the cost savings derived from provisioning SaaS services are being reinvested into developing new skillsets and new strategic technological priorities and growth areas such as Big Data, security, analytics and mobility.

3.5   CONCLUSION

Using an in-depth case study approach incorporating two incumbent IT service providers, this study reveals six tangible payoffs which have manifested across their core business model's domains. These payoffs are categorised as being economic, business and transformative. We now enumerate the implications of the following study. First, we used the STOF business model framework as an analytical anchor in order to present a general understanding of the transformative, beneficial and constraining impacts of SaaS-based digital transformation on incumbent IT service providers. Such IT-based business model insights, *"prepare the path for significant contributions in understanding underlying mechanisms of business model success and failure"* (Veit et al. 2014, p. 50) in an increasingly digitised enterprise world. Moreover, this research provides much needed insights into how SaaS-based digital transformation changes entire business models (Benlian et al. 2016). Second, this study demonstrates that both case organisations are reaping top line (e.g. increased organisational agility, increased sales, enhanced technological capabilities), and bottom line (e.g. reduced operational costs, improved customer experience and satisfaction) payoffs. Thus, this study can serve as a baseline for entrenched incumbent IT service providers to juxtapose and weigh up the transformative, business and economic payoffs that can be derived from provisioning SaaS technologies. It was interesting to note that both case organisations also experienced specific constraints (organisational and SaaS technology level) which inhibited both case organisations' ability to effectively leverage the payoffs of SaaS-based business models. Future research could investigate how IT service providers develop workarounds in order to overcome these constraints. Additionally, our study specifically focused on incumbent IT service providers. Future studies could broaden this scope and investigate how SaaS impacts the business models of IT service providers who were 'born on the cloud'. It is also interesting to further examine the difference between those IT services providers who failed to successfully leverage the business value of SaaS technologies versus those who did reap the rewards. Finally, other creditable business model frameworks such as the business model canvas may also provide additional insights and deserve further exploration as well.

References

Al-Debei, M. M., & Fitzgerald, G. (2010, March). The Design and Engineering of Mobile Data Services: Developing an Ontology Based on Business Model Thinking. In *IFIP Working Conference on Human Benefit through the Diffusion of Information Systems Design Science Research*, 28–51. Berlin, Heidelberg: Springer.

Benlian, Alexander, W. Kettinger, Ali Sunyaev, and T. Winkler. 2016. The Transformative Value of Cloud Computing. *Journal of Management Information Systems* 35: 719–739.

Bouwman, Harry, Edward Faber, Timber Haaker, Björn Kijl, and Mark De Reuver. 2008. Conceptualizing the STOF Model. In *Mobile Service Innovation and Business Models*, 31–70. Berlin, Heidelberg: Springer.

Clohessy, Trevor, Thomas Acton, and Lorraine Morgan. 2017. The Impact of Cloud-based Digital Transformation on IT Service Providers: Evidence from Focus Groups. *International Journal of Cloud Applications and Computing (IJCAC)* 7 (4): 1–19.

Díaz-Díaz, Raimundo, Luis Muñoz, and Daniel Pérez-González. 2017. Business Model Analysis of Public Services Operating in the Smart City Ecosystem: The Case of SmartSantander. *Future Generation Computer Systems* 76: 198–214.

Flyvbjerg, Bent. 2006. Five Misunderstandings about Case-Study Research. *Qualitative Inquiry* 12 (2): 219–245.

Gartner. 2018. Gartner Forecasts Worldwide Public Cloud Revenue to Grow 21.4 Percent in 2018. Accessed 1 February 2018. https://www.gartner.com/newsroom/id/3871416.

Glaser, Barney G., and Anselm L. Strauss. 1967. *The Discovery of Grounded Theory: Strategies for Qualitative Theory*. New Brunswick, NJ: Aldine Transaction.

Guo, Zhiling, and Dan Ma. 2018. A Model of Competition between Perpetual Software and Software as a Service. *MIS Quarterly* 42 (1): 1.

Henfridsson, Ola. 2014. The Power of an Intellectual Account: Developing Stories of the Digital Age. *Journal of Information Technology* 29: 356–357.

Hess, Thomas, Christian Matt, Alexander Benlian, and Florian Wiesböck. 2016. Options for Formulating a Digital Transformation Strategy. *MIS Quarterly Executive* 15 (2): 6.

Iyer, Bala, and John C. Henderson. 2012. Business Value from Clouds: Learning from Users. *MIS Quarterly Executive* 11 (1): 51.

Klein, Heinz K., and Michael D. Myers. 1999. A Set of Principles for Conducting and Evaluating Interpretive Field Studies in Information Systems. *MIS Quarterly* 23: 67–93.

Lee, Y. Y., Noridah, N., Hassan, S. A. A. S., & Menon, J. (2014). *Absence of Helicobacter Pylori is not Protective Against Peptic Ulcer Bleeding in Elderly on Offending Agents: Lessons from an Exceptionally Low Prevalence Population.* PeerJ, 2, e257.

Osterwalder, A. and Pigneur, Y., 2010. Business Model Generation: A Handbook for Visionaries. *Game Changers, and Challengers.* John Wiley & Sons.

Peters, Christoph, Ivo Blohm, and Jan Marco Leimeister. 2015. Anatomy of Successful Business Models for Complex Services: Insights from the Telemedicine Field. *Journal of Management Information Systems* 32 (3): 75–104.

Van Kerkhoff, L. (2014). Developing Integrative Research for Sustainability Science through a Complexity Principles-based Approach. *Sustainability Science* 9: 143–155.

Veit, Daniel, Eric Clemons, Alexander Benlian, Peter Buxmann, Thomas Hess, Dennis Kundisch, Jan Marco Leimeister, Peter Loos, and Martin Spann. 2014. Business Models. *Business & Information Systems Engineering* 6 (1): 45–53.

Wirtz, H. (2011). Innovation Networks in Logistics-management and Competitive Advantages. *International Journal of Innovation Science,* 3(4): 177–192.

Wirtz, B. W., Pistoia, A., Ullrich, S., & Göttel, V. (2016). Business models: Origin, Development and Future Research Perspectives. *Long Range Planning,* 49(1): 36–54.

Yin, Robert K. 2003. *Case Study Research Design and Methods,* Applied Social Research Methods Series 5. 3rd ed. SAGE.

CHAPTER 4

Cloud Service Brokerage: Exploring Characteristics and Benefits of B2B Cloud Marketplaces

Victoria Paulsson, Vincent C. Emeakaroha, John Morrison, and Theo Lynn

Abstract With the increasing popularity of cloud computing, a new technology and business model called cloud service brokerage (CSB) is emerging. CSB is, in essence, a middleman in the cloud-computing supply chain to connect prospective cloud buyers with suitable service providers. This chapter focuses on a type of CSB, B2B cloud marketplaces. Recently,

V. Paulsson (✉)
Irish Centre for Cloud Computing and Commerce, Dublin, Ireland

V. C. Emeakaroha
Cork Institute of Technology, Cork, Ireland

J. Morrison
University College Cork, Cork, Ireland

T. Lynn
Irish Institute of Digital Business, DCU Business School, Dublin, Ireland

© The Author(s) 2020
T. Lynn et al. (eds.), *Measuring the Business Value of Cloud Computing*, Palgrave Studies in Digital Business & Enabling Technologies, https://doi.org/10.1007/978-3-030-43198-3_4

this type of marketplace has evolved into two broad categories—business application marketplaces and API marketplaces. This chapter reviews the characteristics of B2B cloud marketplaces, and their benefits, which include ease-of-use and ease-of-integration, enhanced security, increased manageability, faster implementation, and cost reduction. The chapter concludes with two mini-case studies, on Salesforce AppExchange and RapidAPI, to illustrate how firms could use B2B cloud marketplaces to generate, capture and measure business value.

Keywords Cloud service brokerage • Application marketplace • Case study • API Marketplace

4.1   Introduction

With the increasing popularity of cloud computing, there is a proliferation of cloud service offerings in the market (Elhabbash et al. 2019). A 2017 report on cloud computing (Columbus 2017) finds that 82 per cent of enterprises are already running projects and applications in the cloud. These businesses are focusing on improving and expanding their use of cloud resources and looking for new ways to gain maximum benefits from the cloud. Attitudes amongst enterprises have improved significantly in the last decade as concerns about cloud security, performance, and available expertise have lessened (Columbus 2017). These positive attitudes toward the cloud in business organisations of various shapes and sizes could result in a higher adoption of business-to-business (B2B) cloud application and API marketplaces (MacInnes 2017).

B2B cloud marketplaces are a type of cloud service brokerage (CSB), an intermediary in the cloud computing value chain. Its main role is to connect prospective cloud users, either individuals or firms, with suitable cloud service providers. While there are CSB definitions available from practitioner literatures, such as Gartner (Lheureux et al. 2012), and MarketsandMarkets (2015), there is no widely accepted definition available in academic literature. This paper subscribes to the definition from the US National Institute of Standards and Technology (NIST) that a CSB is "an entity that manages the use, performance, and delivery of cloud services, and negotiates relationships between Cloud Providers and Cloud Consumers" (Sill et al. 2013). Accordingly, in this chapter, CSB refers to a business model and a technology, whose main purpose is to connect prospective cloud service customers with cloud service providers in the cloud.

This chapter aims to provide some insight in to the structure of the CSB landscape, and explore the functional characteristics and benefits of B2B cloud marketplaces. We illustrate these with two mini-case studies on Salesforce AppExchange and RapidAPI.

4.2   The Four Tiers of Cloud Service Brokerage

Before we delve deeper into the topic of B2B cloud marketplaces, it is necessary to build a common understanding of the overall structure of the CSB landscape, and where B2B cloud marketplaces are situated. Fowley et al. (2013) categorise CSB from an architecture platform perspective. They identify three platforms: cloud management (Tier 1), cloud broker (Tier 2) and cloud marketplace (Tier 3) platforms. We extend this categorisation by adding a fourth platform, a cloud marketplace enablement platform (Tier 4), as the industry is expanding in this direction since Fowley et al.'s publication. A brief discussion of the these tiers will be useful in distinguishing cloud marketplaces from other CSB tiers.

Tier 1: Cloud management platform—a cloud architecture platform which offers the complete lifecycle of cloud-related management activities such as the design, deployment and provisioning of cloud infrastructure including advanced monitoring of cloud resource utilisation. An exemplar Tier 1 platform use case is a *management platform for heterogeneous distributed data centres*. These platforms run virtual infrastructures on top of hardware to build private, public or hybrid cloud solutions, such as OpenNebula and Eucalyptus.

Tier 2: Cloud broker platform—supports value added brokering activities such as aggregation, integration, and customisation. These activities require a specific language to make several applications work together in a uniform manner. There are many examples of CSB that specialize in these three core value generators of cloud broker platforms (Lheureux et al. 2012). First, *aggregation CSBs* bring together multiple services at scale e.g. single sign-on, unified billing, chargeback and show back. Examples of aggregation CSBs include BlueWolf and CloudNation. Second, *integration CSBs* focus on making multiple clouds work together in an integrated manner e.g. Dell Boomi and HPE Helion. Lastly, *customisation*

CSBs build new features and functions on top of existing cloud applications according to business needs e.g. LTech.

Tier 3: Cloud marketplace platform—here a CSB builds upon a cloud broker platform to provide a marketplace, which brings providers and consumers together. Two key features in the cloud marketplace are (i) service descriptions for core and integrated services, and (ii) trust. Two types of cloud marketplace are present in the market at the moment: (1) *Business application marketplaces* and (2) *Application programming interface (API) marketplaces.* Business application marketplaces provide cloud application catalogues to perspective business buyers. Examples from the private sector are Microsoft Azure Marketplace, Salesforce AppExchange, Google Apps Marketplace, and the Amazon Web Service (AWS) Marketplace. Examples from the public sector are the Gov.UK Digital Marketplace from the UK government, and the Federal Risk and Authorisation Management Program (FedRAMP) from the US Government. More recently, APIs have emerged as a fundamental building block in the digital economy, connecting organisations, technologies and data. If data is the new oil, APIs are the new pipelines. An API is a set of functions and procedures allowing the creation of applications that access the features or data of a system, application, or other service. They increasingly play an important role in the interoperability of systems, whether cloud-based or otherwise, both internally and externally, and the exchange of data. APIs are bidirectional—they can provide and consume—and can be public (open) or private. A key characteristic of APIs is their abstraction from systems and infrastructure thus allowing third parties to build applications and services that consume APIs. API marketplaces allow cloud software vendors to buy, sell, and/or exchange APIs. Examples include RapidAPI and Apiculture.

Tier 4: Cloud marketplace enablement platform—a multi-tenant platform with a cloud-enabled business model that assists companies to create their own cloud marketplaces. The platform architecture is developed in such a way that it can be licensed, reused and customised to fit many business contexts. Providers of cloud marketplace enablement platforms typically offer a fully integrated platform with a range of value-added services. An exemplar CSB use case in this tier is *Marketplace-as-a-service (MaaS)*. MaaS is a business model in which a MaaS-oriented firm develops a general cloud marketplace platform and white labels this platform to businesses interested in offering a cloud marketplace as part of their business (Fischer 2012). MaaS clients do not have to reinvent the wheel on the technological side of building and managing cloud marketplace infrastructure, so they can focus on strategic issues around marketplaces. AppDirect is an example of a MaaS vendor.

This paper focuses only on Tier 3, B2B cloud application marketplace platforms. The next section will delve into its characteristics.

4.3 CHARACTERISTICS OF B2B CLOUD MARKETPLACE PLATFORMS

B2B cloud marketplaces are a form of *multisided platform business model,* in which two or more parties, for example customers and vendors etc., have a direct interaction with one another through the platform (Brokaw 2014). The platform acts as a middleman between buyers and sellers. It earns a commission and/or revenue sharing for transactions taken place through the platform. The marketplaces *aggregate* a large selection of cloud services, be it business applications or APIs, from multiple software vendors, and offer an integrated service catalogue to cloud customers (Cantara 2015). Aggregation is the core business value that these marketplaces offer, but they could expand to generate value in other ways e.g. integration and customisation. Customers can search for suitable cloud offerings by either browsing through standard categories suggested by the marketplace operators or searching through service descriptions, which accompany every cloud offering (MarketsandMarkets 2015). A successful application or API marketplace attracts a critical mass of customers and software vendors to build economies of scale (Brokaw 2014). Software vendors benefit from access to a large pool of customers which makes an investment to develop new applications and APIs worthwhile. When development costs are spread thinner among a large pool of customers, profits are more likely to increase. Customers also benefit from accessing a large collection of software applications and APIs that could meet their business requirements (Brokaw 2014).

B2B cloud marketplaces offer a wide range of value-added services to the various marketplace actors. First, B2B cloud marketplaces offer many fundamental supports to both sellers and buyers alike. For software vendors, *product planning and development,* and *sales and marketing* are some of the areas that marketplace operators offer an invaluable support. Many B2B cloud marketplace platforms run a partner program with software vendors to help them with the application planning and development side e.g. the AWS Partner Network, and the Salesforce AppExchange Partner Success program. These partner programmes promote new and existing software vendors to further fine-tune their offerings to fit customer requirements (Gill et al. 2016). Requirements to participate in these programmes depend on a given marketplace operator's policy. A general rule of thumb is that the higher the sales revenue that a cloud application or API generates, the more support it tends to receive from marketplace operators. These partner programmes are beneficial for all parties involved. The software vendors are more confident that their applications and APIs are compatible with what the market is looking for. Cloud service customers are satisfied with cloud solutions they purchase and, as a result, might expand their cloud application and API use into other areas. Cloud marketplace operators earn more brokerage fees as applications and APIs become more popular, and as marketplaces attract more business transactions. These are all positive reinforcements for a marketplace business cycle. Partner portals in marketplaces typically provide vendors with key data, insights, indicators and reports including sales leads, conversion rates, customer feedback, and service usage (Oracle 2015; AWS 2016; Salesforce 2016).

For cloud customers, using a cloud marketplace provides them with two core value-added services: (1) *Single-sign-on (SSO)* and (2) *Integrated event and billing management.* Marketplace operators usually provide customers with an ability to use a single username and password to sign onto multiple applications and APIs (AppDirect 2013). SSO has a number of key benefits for firms and their users. First, SSO is associated with increased productivity and improved security. Users enjoy a single point of access to all applications, APIs and resources that they need to get their work done in one convenient portal; they simply log in once to get access to everything they need. They no longer have to waste time finding and logging into separate applications (Kortright 2018). Second, SSO authentication improves overall security from both a system architecture and a behavioural perspective. Architecturally speaking, the SSO authentication model enhances the overall security because security credentials are accepted and processed only at a specific SSO server; no security credentials are transmitted to other systems (Mecca et al. 2016). Behavioural-wise, maintaining multiple passwords for different access points requires a high level of cognitive effort from users. Therefore, users tend to engage in many poor password practices, for example writing down passwords, repeated passwords, and creating a commonly used password for all access points

(Kortright 2018). SSO minimises risk from poor password practices. *Integrated event and billing management* involves marketplace operators providing an integrated portal to monitor and manage applications in use. The portal should be sophisticated enough to address and report complex use, billing and monitoring situations such as billing relationships between users, service level and billing intervals, introductory pricing, one-time fees, free trials, upgrades and renewals (Microsoft Azure 2016a; AppDirect 2013). The portal should also allow administrators to view all costs, manage, edit and/or cancel existing subscriptions.

Actors on both sides of the cloud marketplace platforms, software vendors and customers, concurrently weigh the payoffs for conducting business through marketplaces against the external risks inherent in it. Therefore, apart from providing the key value proposition as a cloud application aggregator and value-added service provider, successful cloud marketplace operators must build and maintain their reputation in the following areas to lower perceived external risk levels: (1) *financial viability*, (2) *corporate governance*, (3) *security and privacy*. First, from a financial viability perspective, marketplace vendors should have a solid financial background and a credible reputation in the market. Cloud customers trust the marketplace vendors to manage their access, through SSO, to relevant applications and APIs, as well as business data. A strong financial background provides an assurance that the cloud service will not be interrupted from business discontinuity (Microsoft Azure 2016a). Second, in terms of corporate *governance*, it is logical to assume that customers are much more likely to conduct a business with cloud marketplace vendors with a strong reputation for corporate governance (Achim et al. 2016). Software vendors and cloud customers, likewise, rely on vendor profiles from reputable sources like Gartner's Magic Quadrant, public ratings and reviews to ensure that their business vision, practices and corporate governance policies are sound (Carraway et al. 2015). Last with regards to security and privacy, vendors should have transparent and compliant data security and privacy policies. Customers should be able to trust that their data and any information processed via the marketplace are secure. Any compromise in this area could seriously undermine the entire viability of cloud marketplace ecosystem since customers on both ends, software vendors and cloud customers, tend to be sensitive to this issue (Sen 2015). Cloud customers are advised to consult service level agreements (SLA) in relation to this matter. Usually marketplace operators operate under a grand scheme SLA. For example, the FedRAMP Marketplace, a cloud marketplace for US federal agencies, ensures that applications listed comply with the US Government's requirements for security in the cloud (DOD 2014). Only applications that have been approved "FedRAMP Ready" are authorised to be listed on the FedRAMP Marketplace (FedRAMP 2019). However, it is possible that specific SLA terms and conditions proffered by certain application vendors may differ from that of the grand-scheme's. Therefore, firms with a strict policy on security and privacy should consult SLAs from both the marketplace operator and vendor to ensure that they are in line with their own policy.

4.4   Benefits of B2B Cloud Marketplace

Based on the characteristics of B2B cloud marketplaces explained above, B2B cloud marketplaces offer a number of advantages to customers. First, *ease-of-use and ease-of-integration* are key advantages that B2B cloud marketplaces offer to cloud customers. As explained earlier, cloud customers can select a cloud application or API that they deem suitable to their business needs, and instantly install it or access it from the cloud. These applications and APIs are typically designed to be intuitive to use or require a bare minimum level of training. Where training is required, software vendors usually provide a wide range of multimedia information, especially video tutorials, to guide users. Users do not have to worry about system upgrades and/or maintenance since software vendors simultaneously and automatically update all underlying software as long as a subscription period is valid (AppDirect 2013). As all cloud marketplace applications and APIs, by definition, are pre-programmed for integration with existing applications, customers can be assured that their newly-acquired applications and APIs will integrate with existing systems (Morin et al. 2012).

Second, *enhanced security* is an advantage of software licensed through cloud marketplaces. In addition to SSO, cloud marketplace vendors can leverage a world-class level of security provided by a team of security professionals at a fraction of the actual costs incurred. Studies suggest storing data on the cloud is much safer than an on-premise alternative. Key security-related issues such as access control, access authentication, data encryption, firewalls, logs and audit trails are better managed and controlled in the cloud (DPC 2015). These resources are not often within reach for an on-premise data storage solution, simply because of the high cost it entails.

Increased manageability is the third benefit of sourcing software services through cloud marketplaces. Traditionally, a software license is for a specific version of software for an indefinite period of time in exchange for a specific sum of revenue. Software vendors earned a new wave of revenue by issuing an upgrade. An investment in new software usually involved a large sum of money once other costs are factored in on top of the license cost, for example implementation, customisation and training costs (Malhotra and Majchrzak 2012). Despite rapid changing business environments and shifting customer requirements, many firms find themselves locked-in with outdated software that are no longer suitable for a business purpose, simply because the management cannot justify investing in an upgrade or new software. As the cloud marketplace business model assumes low customer switching costs, software vendors need to ensure customer satisfaction and usage. In comparison to the old licensing model, customers hold an upper hand as they are able to add, remove and modify services whenever they want (Morin et al. 2012). Cloud customers have greater flexibility to manage their software needs as their market and customers' requirements shift, through the integrated event and billing management portal discussed earlier. They can easily scale up, scale down, or switch to a new vendor.

Fourth, *faster implementation* is another clear benefit from sourcing through a cloud marketplace. Software available on cloud marketplaces typically have programmed interfaces that allow business technologists, i.e. business users who are IT competent, to integrate the applications with existing IT systems (Oracle 2012). Expensive scarce IT professionals and consultants are typically not needed to install these applications, and if needed, the amount of time and effort is significantly reduced. New software should be ready for use after a few installation clicks. For example, GetFeedback (www.getfeedback.com) is a survey application listed on Salesforce AppExchange. It allows users to send surveys to customers, while the survey results will be autonomically synced into Salesforce CRM solution. Its main selling points, over other non-AppExchange survey applications, are ease of installation and a seamless integration capability with Salesforce so that any data received will be directly linked to the Salesforce CRM database.

The last benefit is *cost reduction*. Cloud marketplace customers can pay for software using a wide range of pricing models, e.g. per user, per month, per hour of usage and per amount of data ingested. These pricing models vary by cloud marketplace and software types. However, a basic principle remains constant: customers only pay for the portion of services that they use. For example, if there is only one person using the software for a given period of time, customers can subscribe to one seat for that period and no further. This contrasts with the old software licensing model, in which customers have to purchase a software license regardless of how much and how often they might use the software. In addition, software on cloud marketplaces usually offer a free trial period to allow customers to decide if the software is right for them (Phelan 2015). This reduces upfront costs for customers and provides cashflow advantages as firms scale up and down.

The next two sections present mini-case studies on two B2B cloud marketplaces. The first case is Salesforce AppExchange, a B2B cloud marketplace offering both applications and APIs, and the second is RapidAPI, a B2B API marketplace. Both mini-case studies focus on how customers could use B2B cloud marketplaces to generate, capture and measure business value.

4.5   Mini-case Study I: Salesforce AppExchange

Salesforce AppExchange is a B2B cloud application and API marketplace for Salesforce customers, developers and partners. All applications listed on AppExchange are certified as compatible with Salesforce.com and pre-integrated with Salesforce.com. There are thousands of applications available on the marketplace for a wide variety of use cases including sales, marketing, integration, customer service, manufacturing, analytics, and back office administration. Pricing models also vary including free, paid and discounted for non-profit.

Localytics (www.localytics.com) is a software vendor listed on Salesforce AppExchange. Localytics has been an App Innovation Partner with the Salesforce Marketing Cloud since 2016. Their software enables Salesforce.com and Localytics customers leverage mobile user data to create a closed loop system with regard to a customer across channels. Initially, Localytics' application combined the power of *geofencing*—the use of GPS or RFID technology to create a virtual geographic boundary to trigger a software to responsed when a mobile device is entering a particular area— with Salesforce.com. Firms can leverage Salesforce.com and Localytics to extract value from and generate new value for customers. For example, Priceline.com uses this combination to send targeted location-based marketing messages to its customers (Localytics 2019). These marketing messages are directed at customers who travel without any advance hotel and/or car-rental bookings. To capture these customers, the marketing messages focus on one-day special offers for hotels and car rentals. A combination of two powerful technologies, geofencing and CRM, with clever marketing messages captures business value for Priceline by allowing the firm to trigger a response to its customers' purchase decisions in real time. The business value from using the Localytics application can be conveniently measured in Salesforce.com through the conversion rate of customers who receive a message and make a booking within the same day, as well as the raw dollar amount received for the bookings. Today, Localytics is offered through AppExchange as an API for as little as US$1 per customer per year. Salesforce.com, Localytics and their joint customers share from combinatory innovation, scale, convenience, and reduced costs.

4.6   Mini-case Study II: RapidAPI

RapidAPI is the largest API marketplace and is a merger between RapidAPI and the former Meshape API marketplace. Firms looking for standard APIs can subscribe to API services available on the marketplace to achieve their business purposes without reinventing the wheel. There are three pricing plans for APIs: free, freemium and paid. APIs are made available by categories such as storage, logistics, database, and search etc. In comparison to a traditional approach to build and maintain one's own APIs, using the RapidAPI marketplace allows firms to accelerate go-to-market, innovate, and generate business value much faster. API marketplaces allow firms to focus on what they do best, whether that is designing a cloud, web or mobile application to meet their target market. They can source relevant APIs through the RapidAPI marketplace and thereby significantly reduce an application/website development time cycle. The APIs are designed to meet the requirements of RapidAPI's marketplace including customer support; vendors can be removed from the marketplace for poor quality or service.

In the Irish Institute for Digital Business, we recently worked with a startup, which for the purposes of this case study, we shall call RecipeApp. The client had no technical background but extensive experience in hospitality. They wanted to build an application that provided (i) recipe data including ingredients, calories and portion sizes, and (ii) the ability to use this data for menu planning, budgeting and procurement, by small business such as delicatessens, coffee shops etc. Instead of investing in a significant effort, in terms of cost, time and human resources, to research and compile a database of recipes, ingredients and nutrition data, they were able to leverage the *Recipe—Food—Nutrition* API from Spoonacular available on the RapidAPI marketplace to provide access to over 365,000 healthy recipes, comprising over 2,600 ingredients, and 115,000 menu items. This API provided a wide range of data including, nutrition analysis, indicate cost breakdown, cooking tips, related recipes, scaling/converting, wine pairings and much more. Not only was the client able to access the data they needed but the additional data inspired additional features and functionality for RecipeApp. Furthermore, from the time of the decision to test Spoonacular, the API was implemented through RapidAPI within an hour at no cost. Indeed the first 150 API calls were free, ideal for testing, after which cost per call started at US$0.002 per point up to 10,000 points per day. For a startup, this predictable and scalable pricing was critical. Using an off-the-shelf API through a cloud marketplace enabled the startup to acclerate their time to market, innovate and demonstrate their proof-of-concept in a fraction of the cost, time and effort if they were to undertake all the data collection or API development themselves.

4.7   Conclusion

This chapter explores B2B cloud marketplaces at both structural and functional levels. From a structural perspective, B2B cloud marketplaces can be classified as a type of CSB (Fowley et al. 2013). At a functional level, firms can generate and capture business value from such marketplaces in a variety of ways depending on their role in the marketplace, be it a marketplace operator, a software vendor, or a customer. While the public is very familiar with consumer app marketplaces like Apple AppStore or Google Play, the general public, business decision makers and indeed, scholars, may be less familiar with cloud marketplaces designed specifically for businesses. In particular, we highlight the emergence of API marketplaces both in their own right and as part of what was traditionally called cloud application marketplaces. We illustrate these with two mini-case studies on Salesforce AppExchange and RapidAPI. There is a paucity of business research in this area. As such, we encourage further research not only on business value research but the antecedents and consequences of participation in cloud marketplaces for different actors and sectors. Similarly, we call for business research on other cloud service brokerage tiers.

References

Achim, Monica-Violeta, Sorin-Nicolae Borlea, and Codruţa Mare. 2016. Corporate Governance and Business Performance: Evidence for the Romanian Economy. *Journal of Business Economics and Management* 17 (3): 458–474.

Adabi, Sepideh, Mozhgan Mosadeghi, and Samaneh Yazdani. 2018. A Real-World Inspired Multi-strategy based Negotiating System for Cloud Service Market. *Journal of Cloud Computing* 7 (1): 17.

AppDirect. 2013. Putting the Cloud Within Reach. http://go.appdirect.com/ hs-fs/hub/390882/file-1311910432-pdf/Collateral/Whitepaper-_Putting_ the_Cloud_Within_Reach.pdf.

AWS. 2016. AWS Marketplace Product Support Connection Onboarding and Seller Guide. Accessed 15 January 2017. https://s3.amazonaws.com/awsmp-loadforms/AWS-Marketplace-Product-Support-Connection-Onboarding-and-Seller-Guide.pdf.

Brokaw, Lisa. 2014. How to Win With a Multisided Platform Business Model. https://sloanreview.mit.edu/article/how-to-win-with-a-multisided-platform-business-model/.

Cantara, Michele. 2015. Four Best Practices for Customization Brokerage of Business Processes in the Cloud. Accessed 27 April 2016. https://www.gartner.com/doc/3106718/best-practices-customization-brokerage-business.

Carraway, Debbie, Everette Allen, Charles Cline, Bill Coker, Leslie Dare, Danny D. Davis, Dan Deter et al. 2015. Adopting Cloud Services at NC State: Guidelines and Considerations. Accessed 22 December 2016. https://oit. ncsu.edu/campus-it/cloud-services-at-nc-state/.

Columbus, Louis. 2017. RightScale 2017 State of the Cloud Report: Azure Gaining in Enterprises. https://www.forbes.com/sites/louiscolumbus/2017/02/18/rightscale-2017-state-of-the-cloud-report-azure-gaining-in-enterprises/#415baf4b8481.

DOD. 2014. Department of Defense Cloud Computing Security Requirements Guide. Accessed 22 December 2016. http://iase.disa.mil/cloud_security/ Documents/u_cloud_srg_v1r0-36.pdf.

DPC. 2015. Data Security Guidance. https://www.dataprotection.ie/viewdoc. asp?DocID=1091. Accessed 30 September 2015.

Elhabbash, Abdessalam, Faiza Samreen, James Hadley, and Yehia Elkhatib. 2019. Cloud Brokerage: A Systematic Survey. *ACM Computing Surveys (CSUR)* 51 (6): 119.

Evans, David S., and Richard Schmalensee. 2005. *The Industrial Organization of Markets with Two-Sided Platforms*. National Bureau of Economic Research.

FedRAMP. 2019. Pursuing a FedRAMP Ready Designation. https://www. fedramp.gov/pursuing-a-fedramp-ready-designation/.

Fischer, Juli. 2012. MaaS (Marketplace-as-a-Service). Accessed 28 June 2016. http://info.appdirect.com/resources/knowledge/marketplace-as-a-service.

Fowley, Frank, Claus Pahl, and Li Zhang. 2013. *A Comparison Framework and Review of Service Brokerage Solutions for Cloud Architectures.* Service-Oriented Computing–ICSOC 2013 Workshops. Springer.

Gill, Nivi, Sean Hogan, and Mike Harvath. 2016. How To Build Badass Cloud Companies—By Avoiding 10 Common Mistakes. Accessed 3 January 2017. https://www.salesforce.com/video/300361/.

Kortright, Scott. 2018. Benefits of Single Sign-on Solutions. http://blog.identityautomation.com/benefits-of-single-sign-on-solutions.

Lheureux, Benoit, Daryl C. Plummer, Neil Wynne, Bill Caffery, James D. Buckner, and Ian Baeazley. 2012. A CIO Primer on Cloud Services Brokerage. Accessed 27 April 2016. https://www.gartner.com/doc/2201415/cio-primer-cloud-services-brokerage.

Localytics. 2019. priceline.com https://appexchange.salesforce.com/servlet/servlet.FileDownload?file=00P3A00000a0r8BUAQ.

MacInnes, Billy. 2017. Working Out What Makes a Good Cloud Marketplace. https://www.computerweekly.com/microscope/news/450420067/Working-out-what-makes-a-good-cloud-marketplace.

Malhotra, Arvind, and Ann Majchrzak. 2012. How Virtual Teams Use Their Virtual Workspace to Coordinate Knowledge. *ACM Transactions on Management Information Systems (TMIS)* 3 (1): 6.

MarketsandMarkets. 2015. Cloud Services Brokerage Market by Types—Global Forecast to 2020. Accessed 15 February 2016. http://www.researchandmarkets.com/research/xwbvvz/cloud_services.

Mecca, Giansalvatore, Michele Santomauro, Donatello Santoro, and Enzo Veltri. 2016. On Federated Single Sign-On in e-Government Interoperability Frameworks. *International Journal of Electronic Governance* 8 (1): 6–21.

Microsoft Azure. 2016a. How Do I Choose a Cloud Service Provider?. Accessed 21 December 2016. https://azure.microsoft.com/en-us/overview/choosing-a-cloud-service-provider/.

———. 2016b. How to Publish and Manage an Offer in the Azure Marketplace. Accessed 2 January 2017. https://docs.microsoft.com/en-us/azure/marketplace-publishing/marketplace-publishing-getting-started#supported-types-of-solutions.

Morin, Jean-Henry, Jocelyn Aubert, and Benjamin Gateau. 2012. *Towards Cloud Computing SLA Risk Management: Issues and Challenges.* 2012 45th Hawaii International Conference on System Sciences. IEEE.

Oracle. 2012. *Ten Questions to Ask Your Cloud Vendor Before Entering the Cloud.* An Oracle White Paper. http://www.oracle.com/us/products/applications/10-questions-for-cloud-vendors-1639601.pdf.

———. 2015. Market Your Oracle Cloud Apps and Services. https://cloud.oracle.com/_downloads/eBook_Marketplace_res2/Oracle_Cloud_Marketplace_Partner_Portal.pdf.

Phelan, Gemma. 2015. Selling Cloud Services on the Digital Marketplace. https://digitalmarketplace.blog.gov.uk/2015/08/28/selling-cloud-services-on-the-digital-marketplace/.

Salesforce. 2016. Salesforce Partner Program Guide for ISVs: A Detailed Guide to the ISV Program. Accessed 15 January 2017. https://partners.salesforce.com/s/ISVPartnerProgramGuidePY2017.pd.

Sen, Jaydip. 2015. Security and Privacy Issues in Cloud Computing. In *Cloud Technology: Concepts, Methodologies, Tools, and Applications*, 1585–1630. IGI Global.

Sill, Alan, Annie Sokol, Craig Lee, David Harper, Eugene Luster, Frederic de Vaulx, Gary Massaferro, and Gilbert Pilz. 2013. NIST Cloud Computing Standards Roadmap. NIST Special Publication 500-291, Version 2. Accessed 13 January 2016. http://www.nist.gov/itl/cloud/upload/NIST_SP-500-291_Version-2_2013_June18_FINAL.pdf.

CHAPTER 5

Refinement of Cost Models for Cloud Deployments through Economic Models Addressing Federated Clouds

Jörn Altmann and Ram Govinda Aryal

Abstract Ten different cloud deployment models (e.g. public clouds, private clouds, and federated clouds) can be selected by users, depending on their requirements for executing their applications. As each of these deployment models comes with certain costs and benefits, they need to be understood by users to ensure they make an optimal selection. In order to support this decision making process, a comprehensive cost model that comprises all relevant cost factors is needed. The aim of this chapter is to present such a comprehensive cost model comprising all relevant cost factors and an economic models underlying the various cloud deployment models. In particular, an architecture for managing federated clouds using an economic model is discussed in the outlook on future research.

J. Altmann (✉) • R. G. Aryal
Technology Management, Economics, and Policy Program, College of
Engineering, Seoul National University, Seoul, South Korea
e-mail: jorn.altmann@acm.org

© The Author(s) 2020
T. Lynn et al. (eds.), *Measuring the Business Value of Cloud
Computing*, Palgrave Studies in Digital Business & Enabling
Technologies, https://doi.org/10.1007/978-3-030-43198-3_5

73

Keywords Cost models • Cloud federation • Value creation
• Deployment models • Economic model • Cloud economics

5.1   Topic Definition

Cloud computing has widely been accepted as an efficient way of using computing resources. It lowers the cost of computing services, provides access to virtually unlimited resources, and allows for flexible charging methods (Jeferry et al. 2015). Nonetheless, there are resource limitations and cost-related shortcomings (Goher et al. 2018). For instance, cloud computing services could still be over-provisioned or under-provisioned (Goiri et al. 2012), which is due to the volatility of demand for computing resources and anti-competitive externalities that exist in the market (Altmann and Kashef 2014; Mohammed et al. 2009). In order to allow cloud providers and cloud customers to deal with this situation, we present a comprehensive cost model comprising relevant cost factors and an economic model underlying the various cloud deployment models.

5.1.1   *Public Clouds, Private Clouds, and Interconnected Clouds*

As cloud users can select among various cloud deployment models, which come with different costs and benefits, an understanding of those costs and benefits is essential for making an optimal selection decision. In order to outline the difference in costs that cloud users face, we distinguish three major types of cloud deployment models: public clouds, private clouds, and interconnected clouds. While public clouds represent clouds that anybody can use as a customer (e.g., Amazon EC2) for a service charge (Altmann et al. 2010), access to a private cloud is limited to the owner of the cloud. Private clouds denote firms' private data centres and host their security-critical services, meeting the firms' computational needs. The data centre is organized using cloud computing technology, as it allows for more efficient organising of the information technology resources of the firm. Public clouds are not necessarily interoperable, as they might have been built with proprietary cloud technologies (Breskovic et al. 2013a, b; Gebregiorgis and Altmann 2015; Maurer et al. 2012). A cloud user, who wants to use several public clouds, would need to use the standards of each cloud. Interconnected clouds are combinations of private clouds and public clouds (Rochwerger et al. 2009), which are achieved technically by making clouds interoperable, giving them the same cloud interfaces. Due to this interoperability, virtual machines can easily be migrated between clouds owned by different cloud providers. It allows users of a cloud to take advantage of the capabilities of other clouds in addition to those of their primary cloud providers (Haile and Altmann 2018). Considering the ownerships, standards, and location of the interconnected clouds, different types of interconnected clouds can be distinguished. A few examples are given in Fig. 5.1.

Overall, eight types of interconnected clouds can be distinguished: public interclouds, private interclouds, hybrid clouds, federated clouds, federated hybrid clouds, hybrid interclouds, federated hybrid interclouds, and federated interclouds.

Distinguishing interconnected clouds from the ownership perspective, clouds can be classified into interclouds and federated clouds. An intercloud is owned by a single provider (e.g., Amazon Web Services), while the clouds of a federated cloud (type 7 in Table 5.1) are owned by several providers. The motivation of interclouds is mainly fault tolerance (e.g., guaranteed availability of customer applications through reliable multi-site deployments (Petcu 2014)) and quality of service (e.g., latency reduction) through a larger computing resource base and their wide geographical

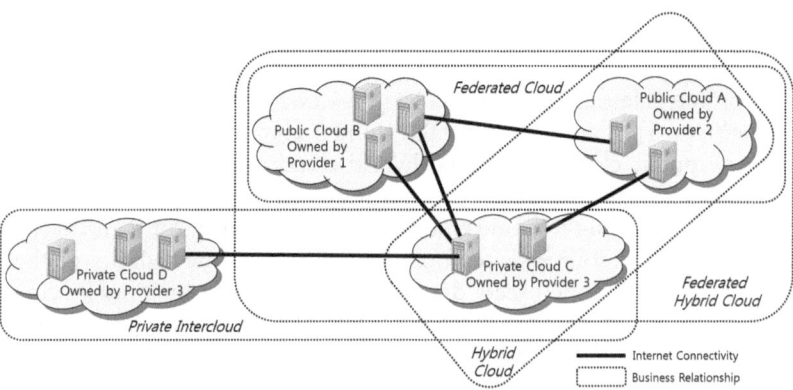

Fig. 5.1 Example of four interconnected clouds (i.e., private intercloud, hybrid cloud, federated hybrid cloud, and federated cloud), being composed of private clouds and public clouds

Table 5.1   Ten types of cloud deployment models

		Ownership			
		Provider X owns one cloud		Provider X owns several clouds	
		Provider gives private access	Provider gives public access	Provider gives private access	Provider gives public access
Cloud standards	Single Cloud Standard is Used by Provider X, which is Different to those of Other Providers.	Private cloud (type 1)	Public cloud (type 2)	Private intercloud (type 4)	Public intercloud (type 5)
	Single Cloud Standard is Used by Provider X, which is Identical to the Standard of one Public Cloud Provider, allowing Interconnection, but Different to those of Other Providers.	Hybrid cloud (type 3)	Public cloud (type 2)	Hybrid intercloud (type 6)	Public intercloud type 5)
	Single Cloud Standard is Used by Provider X, and a Service Level Agreement between Provider X and a few Public Cloud Providers Exists.	Federated hybrid cloud (type 8)	Federated cloud (type 7)	Federated hybrid intercloud (type 9)	Federated intercloud (type 10)

distribution (Hassan et al. 2014). An intercloud can be private (type 4) or public (type 5). If an interconnected cloud is composed of a private cloud and one public cloud, it is called hybrid cloud (type 3). If an interconnected cloud is composed of several private clouds and one public cloud, it is called hybrid intercloud (type 6). It is used by firms, if their demand for computing resources is temporarily in excess of the capacity of their private clouds and the excess demand can be covered by a public cloud (Bossche et al. 2013). With respect to federated clouds, the cloud providers participating in a cloud federation have reached an additional service level agreement, called federation service level agreement (FSLA), for cooperating on deployments of customer applications. Federated clouds enable marketplaces and trading of standardized goods (Altmann et al. 2010; Maurer et al. 2012). They also enable small cloud providers to collaborate and gain access to an increased number of cloud infrastructure resources and to benefit from the economies of scale through the aggregation of both requests and resources (Haile and Altmann 2016a, b; Kim et al. 2014). An interconnected cloud is called federated hybrid cloud (type 8), if the interconnected public clouds signed an FSLA and the private cloud accesses the federated clouds for additional cloud services. The federated hybrid cloud model is an extension to the hybrid cloud model. If the private cloud is an intercloud, the model is called federated hybrid intercloud (type 9). The last model is the federated intercloud (type 10), which does not comprise a private cloud. All ten interconnected cloud deployment models are shown in Table 5.1, which presents them according to cloud ownership, access rights, and cloud standards.

When it comes to the adoption of any of those ten cloud deployment models, companies need to consider various factors including application complexity, available technology options, available support, security provisions and, more importantly, the cost.

5.1.2   The Need for Detailed Cost Models for Clouds

For many companies, the migration of the workload to the public cloud has helped them achieve rapid deployment of their applications and reduce operational costs for their data centres (Rayport and Heyward 2009). A recent article by McKinsey states that a company using a traditional computing model can potentially make savings (both labor and non-labor combined) of 9% by adopting a private cloud, and up to 61% by adopting a public cloud (Gu et al. 2018). Current research already indicates that further reduction of the cost of cloud service provisioning is possible through cloud federations (Altmann and Kashef 2014; Aryal and Altmann 2018; Goiri et al. 2012). However, in order to make informed decisions on a migration to clouds (Kauffman et al. 2018; Khajeh-Hosseini et al. 2012; Truong and Dustdar 2010), details about the costs of clouds are needed. An in-depth cost analysis of the various options, which affect the overall cost of adopting an appropriate cloud deployment model, can help. The cost analysis can comprise calculating different economic values (e.g., the Net-Present-Value, the Return-on-Investment, the Discounted-Payback-Period, and the Benefit-to-Cost-Ratio), which are essential for any business decision (Klems et al. 2008), and, herewith, would enable organisations to determine which cloud deployment model is most beneficial.

Due to its importance for cloud deployment decision making, cost models have gained significant attention of the research community in recent years.

5.2   STATE OF THE ART IN COST MODELS

Reviewing previous literature on cloud cost models, which has been published between year 2005 and year 2019 and was found searching Google Scholar with combinations of search terms 'cost', 'cost model', 'cloud', 'cloud computing', 'grid', 'cost factor' and 'cost-benefit', suggests the need to consider 21 cost factor groups for estimating the cost of IaaS services applicable for different cloud deployment models. The identified cost factors and their classifications are presented in Table 5.2. The classification of the 21 cost factors comprises six main categories (i.e., electric power, system infrastructure, software, human resources, business premises, and cloud services). These cost factor groups are also classified according to the cloud deployment model, which need to consider this group of cost factors.

A detailed description of each of the categories and groups of cost factors is given in the following subsections.

5.2.1   *Cost Factor Category: Electric Power*

Electric power consumption is one of the major factors contributing to the cost of clouds. A cloud consumes power basically for two activities: (1) powering data center devices such as routers, switches, gateways, servers, and storage devices (Alford and Morton 2009; Armbrust et al. 2009; Kondo et al. 2009; Opitz et al. 2008; Patel and Shah 2005; Risch and Altmann 2008; Tak et al. 2011); and (2) operating HVAC cooling system devices (Alford and Morton 2009; Armbrust et al. 2009; Patel and Shah 2005; Opitz et al. 2008; Tak et al. 2011). In order to have the precise estimations, it is necessary to consider the power consumption of each device under different conditions (i.e., when it is running at no load, average load, and full load) (Opitz et al. 2008).

5.2.2   *Cost Factor Category: System Infrastructure*

The cost associated with acquiring a hardware system infrastructure for in-house use is referred to as system infrastructure cost. System

Table 5.2 Cost factors associated with cloud deployment models

Cost factor category	Cost factor group		Target cloud deployment type	Literature on cost factors													
				Risch and Altmann (2008)	Tak et al. (2011)	Kondo et al. (2009)	Armbrust et al. (2009)	Haijat et al. (2010)	Truong and Dustdar (2010)	Alford and Morton (2009)	Khajeh-Hosseini et al. (2012)	Opitz et al. (2008)	Altmann and Robirtana (2010)	Hwang et al. (2013)	Patel and Shah (2005)	Amazon (2019)	Altmann and Kashef (2014)
(a) Electric power	(a1)	Cooling	1, 3, 4, 6, 8, 9									O				O	O
	(a2)	Electronic devices (idle)	1, 3, 4, 6, 8, 9									O				O	O
	(a3)	Electronic devices (use)	1, 3, 4, 6, 8, 9	O	O	O	O			O		O			O	O	O
(b) System infrastructure	(b1)	Server	1, 3, 4, 6, 8, 9	O	O	O				O		O	O			O	O
	(b2)	Network device	1, 3, 4, 6, 8, 9		O					O			O			O	O
(c) Software	(c1)	Basic server software license	All		O					O			O		O	O	O
	(c2)	Middleware license	All									O			O	O	O
	(c3)	Application software license	All		O					O		O	O			O	O
(d) Human Resources	(d1)	Software maintenance	1, 3, 4, 6, 8, 9		O					O					O	O	O
	(d2)	Hardware maintenance	1, 3, 4, 6, 8, 9		O					O		O			O	O	O
	(d3)	Other support	1, 3, 4, 6, 8, 9		O	O				O		O				O	O

(continued)

Table 5.2 (continued)

Cost factor category	Cost factor group	Target cloud deployment type	Literature on cost factors													
			Risch and Altmann (2008)	Tak et al. (2011)	Kondo et al. (2009)	Armbrust et al. (2009)	Hajjat et al. (2010)	Truong and Dustdar (2010)	Alford and Morton (2009)	Khajeh-Hosseini et al. (2012)	Opitz et al. (2008)	Altmann and Rohitratana (2010)	Hwang et al. (2013)	Patel and Shah (2005)	Amazon (2019)	Altmann and Kashef (2014)
(e) Business premises	(e1) Rack, air conditioner	1, 3, 4, 6, 8, 9							O					O		O
	(e2) Cabling	1, 3, 4, 6, 8, 9							O					O		O
	(e3) Facility	1, 3, 4, 6, 8, 9				O								O		O
(f) Cloud services	(f1) Internet connectivity	All			O	O	O									O
	(f2) Cloud server use	2, 3, 5, 6, 7, 8, 9, 10	O	O	O	O	O	O		O			O		O	O
	(f3) Data transfer into cloud	2, 3, 5, 6, 7, 8, 9, 10	O	O	O	O	O	O		O	O		O		O	O
	(f4) Data transfer from cloud	2, 3, 5, 6, 7, 8, 9, 10	O	O	O	O	O	O		O	O		O		O	O
	(f5) Cloud storage use	2, 3, 5, 6, 7, 8, 9, 10	O	O	O	O	O	O		O			O		O	O
	(f6) Data transfer between clouds	4, 5, 6, 7, 8, 9, 10													O	O
	(f7) Number of deployments	4, 5, 6, 7, 8, 9, 10 Interconnected cloud														O

infrastructure cost can be classified into two groups. The first group includes the cost of acquiring computing equipment such as servers and storage devices for in-house use (Alford and Morton 2009; Altmann and Rohitratana 2010; Kondo et al. 2009; Opitz et al. 2008; Risch and Altmann 2008; Tak et al. 2011). The second group includes the cost of acquiring network equipment such as routers (Alford and Morton 2009; Altmann and Rohitratana 2010; Tak et al. 2011). For this cost of system infrastructure, as suggested by Opitz et al. (2008), it is important to consider the time period (i.e., depreciation period), during which this equipment can be used. It is a normal practice for this equipment to have a three-year economic lifetime.

5.2.3 Cost Factor Category: Software

The software cost factor category includes the cost incurred in purchasing software licenses for in-house use. In particular, the operations of data centers include three groups of software licenses. The first category may be referred to as basic server software licenses, which include operating system software licenses and licenses for system administration such as backup system software (Alford and Morton 2009; Altmann and Rohitratana 2010; Patel and Shah 2005; Tak et al. 2011). Similarly, the second group of licenses includes commercial middleware software that is required for operating a cloud (Opitz et al. 2008; Patel and Shah 2005). The last group includes licenses for application software, which provides value directly to customers. An example is a software for enterprise resource planning (Alford and Morton 2009; Altmann and Rohitratana 2010; Opitz et al. 2008; Tak et al. 2011). It is also important to note that depending on the vendor, software may be purchased only under certain licensing policies (e.g., perpetual license, license that require periodic renewal, and license that charge according to the number of end users). Due to these different licensing policies (Altmann and Rohitratana 2010), the cost can vary widely.

5.2.4 Cost Factor Category: Human Resources

The human resources cost category includes salaries to be paid for technicians, who maintain the hardware infrastructure (Alford and Morton 2009; Armbrust et al. 2009; Opitz et al. 2008; Tak et al. 2011), technicians, who maintain software applications (Alford and Morton 2009; Patel and Shah 2005; Tak et al. 2011), and technicians, who provide support services (Alford and Morton 2009; Kondo et al. 2009; Opitz et al. 2008; Tak et al. 2011). Due to the differences in economic conditions, cost of living, and the availability of labor in different countries, this cost category is largely determined by the geographical location of a data center.

5.2.5   Cost Factor Category: Business Premises

The business premises cost category includes the basic costs involved in setting up the basic facilities required for establishing an in-house data centre (private cloud). Important cost factors include (1) the cost of purchasing or leasing the data centre facility (Armbrust et al. 2009; Patel and Shah 2005); (2) the cost of all installations that ensure security and reliability of the data center such as HVAC cooling systems, physical security systems, server racks, and other non-electronic devices (Alford and Morton 2009; Patel and Shah 2005); and (3) the cost for cabling and networking required for the operation of the data centre (Alford and Morton 2009; Patel and Shah 2005).

5.2.6   Cost Factor Category: Cloud Services

This category of cost factors includes all intangible cost items directly related to the use of cloud services. They are classified into seven groups. One of the groups comprises the server usage cost (e.g., per hour price of virtual machine (VM) instance time duration) (Hajjat et al. 2010; Hwang et al. 2013; Khajeh-Hosseini et al. 2012; Kondo et al. 2009; Risch and Altmann 2008; Truong and Dustdar 2010). The second group includes cost items related to the cost of storage (Armbrust et al. 2009; Hajjat et al. 2010; Hwang et al. 2013; Khajeh-Hosseini et al. 2012; Kondo et al. 2009; Opitz et al. 2008; Risch and Altmann 2008; Tak et al. 2011; Truong and Dustdar 2010). The cost of Internet service for the firm is another cost group (Hajjat et al. 2010; Kondo et al. 2009). The cost of data transfers into the cloud and the cost of data transfer out from the cloud make two more groups (Armbrust et al. 2009; Hajjat et al. 2010; Hwang et al. 2013; Khajeh-Hosseini et al. 2012; Kondo et al. 2009; Opitz et al. 2008; Risch and Altmann 2008; Tak et al. 2011; Truong and Dustdar 2010). As another cost factor group, the cost associated with the transfer of data between clouds has been identified (Altmann and Kashef 2014). This cost group has not received much attention in literature, although it is important in the case of interconnected clouds. Its significance can be seen from the fact that Amazon charges higher prices for data transfer between its clouds located in different regions than for data centers located in the same region. The last cost factor group is the deployment cost. This cost group had also not received much attention in previous literature. Although the probability of VM migrations is low in case of hybrid cloud deployment types, for interconnected clouds, this cost factor group can become significant. Various events may trigger new service deployments that are more economically efficient than the existing deployments. Deployment cost may be determined as the number of deployments multiplied by the cost of migration for each deployment.

5.3 AVENUES FOR FUTURE RESEARCH: ECONOMIC MODELS FOR FEDERATED CLOUDS

Interconnected clouds, in particular, cloud federations, have been considered as a way for cloud providers to address the limitations of clouds and to decrease the cost of service provisioning by means of resource aggregations and reliable multi-site deployments. Despite these significant promises, we cannot find any cloud federation in operation in the commercial market. Our review of the cloud federation literature (Aryal and Altmann 2017; Coronado and Altmann 2017; Hassan et al. 2014; Jeferry et al. 2015; Samaan 2014; Wang et al. 2012) identified only a few notable causes. In particular, a lack of proper economic models has been identified as an important hindering factor in the adoption of cloud federations (Breskovic et al. 2011; Haile and Altmann 2015).

These economic models can incentivize cloud providers for forming and operating federations and for sharing revenue. As revenue sharing is directly linked to how resources are shared among federation members (Roth 1988), an efficient solution should specify jointly how members of a cloud federation share the infrastructure resources and the revenue generated through the service provisioning with shared resources. The solution needs also to connect these algorithms for resource sharing (i.e., service placement) and revenue sharing (i.e., business logic) with an accounting system. Appropriate algorithms for revenue sharing are required to incentivize properly the federation members for fulfilling the service requests that they should serve and the service requests that they bring into the federation. It also requires considering a large number of

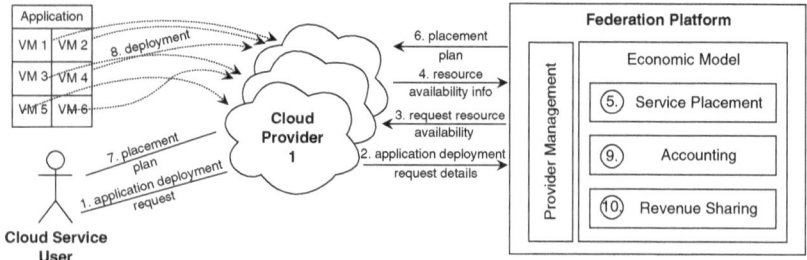

Fig. 5.2   Use case for applying economic models for cloud federations

geographically distributed providers offering heterogeneous services with varying QoS guarantees (Aryal and Altmann 2017; Aryal and Altmann 2018). Despite this interesting challenge, only very few researchers have started working on these economic models related to service placement and revenue sharing in federated clouds and have proposed algorithms (Aryal et al. 2019).

Considering the need for algorithms for service placement and revenue sharing, a system architecture for a cloud federation platform and a use case for applying an economic model in cloud federation deployments can be envisioned as shown in the following figure (Fig. 5.2). The economic model in this architecture is implemented through three modules: service placement, accounting, and revenue sharing.

The use case shown in Fig. 5.2 highlights the workings of the system architecture. It illustrates that a cloud service user in need of application deployment in the cloud initiates an application deployment request (step 1) through a cloud provider (Cloud Provider 1), who is a member of a cloud federation. The cloud federation platform receives the request through the provider management module (step 2) and triggers the service placement module within the economic module for determining the service placement plan based on the available resources (step 3 and step 4) by identifying the best possible combination of the federated cloud resources (step 5 and step 6). The service nodes of the customer application get deployed as per placement plan (step 7 and step 8). The accounting module within the economic model records the transaction and resource consumption in the customer account, using the service provisioning details for the request (step 9). Finally, the revenue sharing module allocates the earned revenue as per the agreed sharing rules (step 10).

As not much research has been performed on the details of economic models for federated clouds, future research is needed to fill this research gap. Researchers should seek to propose economic models for the governance of cloud federations that incentivize cloud providers to form and sustain the operation of cloud federations. The architecture shown in Fig. 5.2 indicates the workings of the modules of such an economic model. These economic models provide guidance for the effective utilization of provider resources in serving customer requests for cloud services and distribute the revenue generated in an appropriate way to the federation members. In particular, they could help in the selection of an optimal set of cloud resources to host application service components and provide a fair, stable, and motivating revenue sharing scheme for cloud federation members with a better return on investments.

References

Alford, Ted, and Gwen Morton. 2009. *The Economics of Cloud Computing: Addressing the Benefits of Infrastructure in the Cloud*. Booz Allen Hamilton.

Altmann, Jörn, and Mohammad Mahdi Kashef. 2014. Cost Model based Service Placement in Federated Hybrid Clouds. *Future Generation Computer Systems* 41: 79–90.

Altmann, Jörn, and Juthasit Rohitratana. 2010. Software Resource Management Considering the Interrelation between Explicit Cost, Energy Consumption, and Implicit Cost. *Multikonferenz Wirtschaftsinformatik* 2010: 71.

Altmann, Jörn, Costas Courcoubetis, and Marcel Risch. 2010. A Marketplace and Its Market Mechanism for Trading Commoditized Computing Resources. *Annals of Telecommunications-annales des télécommunications* 65 (11–12): 653–667.

Amazon. Simple monthly calculator. 2019. Accessed March 2019. https://calculator.s3.amazonaws.com/index.html.

Armbrust, Michael, Armando Fox, Rean Griffith, Anthony D. Joseph, Randy H. Katz, Andrew Konwinski, Gunho Lee, David A. Patterson, Ariel Rabkin, Ion Stoica, Matei Zaharia. 2009. Above the Clouds: A Berkeley View of Cloud Computing. *UCB/EECS*, 2009–2028. EECS Department, UCB.

Aryal, Ram Govinda, Jamie Marshall, and Jörn Altmann. 2019. *Architecture and Business Logic Specification for Dynamic Cloud Federations*. GECON 2019, 16th International Conference on Economics of Grids, Clouds, Systems, and Services, LNCS 11113, pp. 83–96. Springer, Leeds, UK.

Aryal, Ram Govinda, and Jörn Altmann. 2017. *Fairness in Revenue Sharing for Stable Cloud Federations*. International Conference on the Economics of Grids, Clouds, Systems, and Services, 219–232. Cham: Springer.

————. 2018. *Dynamic Application Deployment in Federations of Clouds and Edge Resources Using a Multiobjective Optimization AI Algorithm.* 2018 Third International Conference on Fog and Mobile Edge Computing (FMEC), 147–154. IEEE.

Van den Bossche, Ruben, Kurt Vanmechelen, and Jan Broeckhove. 2013. Online Cost-Efficient Scheduling of Deadline-Constrained Workloads on Hybrid Clouds. *Future Generation Computer Systems* 29 (4): 973–985.

Breskovic, Ivan, Michael Maurer, Vincent C. Emeakaroha, Ivona Brandic, and Jörn Altmann. 2011. *Towards Autonomic Market Management in Cloud Computing Infrastructures.* CLOSER, 24–34.

Breskovic, Ivan, Ivona Brandic, and Jörn Altmann. 2013a. *Maximizing Liquidity in Cloud Markets through Standardization of Computational Resources.* SOSE 2013, International Symposium on Service-Oriented System Engineering, San Jose, USA.

Breskovic, Ivan, Jörn Altmann, and Ivona Brandic. 2013b. Creating Standardized Products for Electronic Markets. *Future Generation Computer Systems*, Elsevier, 29 (4): 1000–1011.

Coronado, Juan Pablo Romero, and Jörn Altmann. 2017. *Model for Incentivizing Cloud Service Federation.* International Conference on the Economics of Grids, Clouds, Systems, and Services, 233–246. Cham: Springer.

Gebregiorgis, Selam Abrham, and Jörn Altmann. 2015. IT Service Platforms: Their Value Creation Model and the Impact of Their Level of Openness on Their Adoption. *Procedia Computer Science* 68: 173–187.

Goher, Syeda ZarAfshan, Peter Bloodsworth, Raihan Ur Rasool, and Richard McClatchey. 2018. Cloud Provider Capacity Augmentation Through Automated Resource Bartering. *Future Generation Computer Systems* 81: 203–218.

Goiri, Íñigo, Jordi Guitart, and Jordi Torres. 2012. Economic Model of a Cloud Provider Operating in a Federated Cloud. *Information Systems Frontiers* 14 (4): 827–843.

Gu, Mark, Krish Krishnakanthan, Anand Mohanrangan, and Brent Smolinski. 2018. The Progressive Cloud: A New Approach to Migration. https://www.mckinsey.com/business-functions/digital-mckinsey/our-insights/the-progressive-cloud-a-new-approach-to-migration.

Haile, Netsanet, and Jörn Altmann. 2015. *Risk-Benefit-Mediated Impact of Determinants on the Adoption of Cloud Federation.* PACIS, 17.

————. 2016a. Value Creation in Software Service Platforms. *Future Generation Computer Systems* 55: 495–509.

————. 2016b. Structural Analysis of Value Creation in Software Service Platforms. *Electronic Markets* 26 (2): 129–142.

————. 2018. Evaluating Investments in Portability and Interoperability between Software Service Platforms. *Future Generation Computer Systems* 78: 224–241.

Hajjat, Mohammad, Xin Sun, Yu-Wei Eric Sung, David Maltz, Sanjay Rao, Kunwadee Sripanidkulchai, and Mohit Tawarmalani. 2010. Cloudward Bound: Planning for Beneficial Migration of Enterprise Applications to the Cloud. *ACM SIGCOMM Computer Communication Review* 41 (4): 243–254.

Hassan, Mohammad Mehedi, M. Shamim Hossain, A.M. Jehad Sarkar, and Eui-Nam Huh. 2014. Cooperative Game-based Distributed Resource Allocation in Horizontal Dynamic Cloud Federation Platform. *Information Systems Frontiers* 16 (4): 523–542.

Hwang, Ren-Hung, Chung-Nan Lee, Yi-Ru Chen, and Da-Jing Zhang-Jian. 2013. Cost Optimization of Elasticity Cloud Resource Subscription Policy. *IEEE Transactions on Services Computing* 7 (4): 561–574.

Jeferry, Keith, George Kousiouris, Dimosthenis Kyriazis, Jörn Altmann, Augusto Ciuffoletti, Ilias Maglogiannis, Paolo Nesi, Bojan Suzic, and Zhiming Zhao. 2015. Challenges Emerging from Future Cloud Application Scenarios. *Procedia Computer Science* 68: 227–237.

Kauffman, Robert J., Dan Ma, and Yu. Martin. 2018. A Metrics Suite of Cloud Computing Adoption Readiness. *Electronic Markets* 28 (1): 11–37.

Khajeh-Hosseini, Ali, David Greenwood, James W. Smith, and Ian Sommerville. 2012. The Cloud Adoption Toolkit: Supporting Cloud Adoption Decisions in the Enterprise. *Software: Practice and Experience* 42 (4): 447–465.

Kim, Kibae, Songhee Kang, and Jörn Altmann. 2014. *Cloud Goliath Versus a Federation of Cloud Davids*. International Conference on Grid Economics and Business Models, 55–66. Cham: Springer.

Klems, Markus, Jens Nimis, and Stefan Tai. 2008. *Do Clouds Compute? A framework for Estimating the Value of Cloud Computing*. Workshop on E-Business, 110–123. Berlin, Heidelberg: Springer.

Kondo, Derrick, Bahman Javadi, Paul Malecot, Franck Cappello, and David P. Anderson. 2009. *Cost-Benefit Analysis of Cloud Computing Versus Desktop Grids*. IPDPS, vol. 9, pp. 1–12.

Maurer, Michael, Vincent C. Emeakaroha, Ivona Brandic, and Jörn Altmann. 2012. Cost–Benefit Analysis of an SLA Mapping Approach for Defining Standardized Cloud Computing Goods. *Future Generation Computer Systems* 28 (1): 39–47.

Mohammed, Ashraf Bany, Jörn Altmann, and Junseok Hwang. 2009. Cloud Computing Value Chains: Understanding Businesses and Value Creation in the Cloud. In *Economic Models and Algorithms for Distributed Systems*, 187–208. Basel: Birkhäuser.

Opitz, Alek, Hartmut König, and Sebastian Szamlewska. 2008. What Does Grid Computing Cost? *Journal of Grid Computing* 6 (4): 385–397.

Patel, Chandrakant D., and Amip J. Shah. 2005. *Cost Model for Planning, Development and Operation of a Data Center*. Palo Alto, CA: Hewlett Packard.

Petcu, Dana. 2014. Consuming Resources and Services from Multiple Clouds. *Journal of Grid Computing* 12 (2): 321–345.

Rayport, Jeffrey F., and Andrew Heyward. 2009. *Envisioning the Cloud: The Next Computing Paradigm and Its Implication for Technology Policy*. White Paper Marketspace LLC. Accessed March 2019. https://marketspacenext.files.wordpress.com/2011/01/envisioning_the_cloud_presentation-deck.pdf.

Risch, Marcel, and Jörn Altmann. 2008. *Cost Analysis of Current Grids and Its Implications for Future Grid Markets*. International Workshop on Grid Economics and Business Models, 13–27. Berlin, Heidelberg: Springer.

Rochwerger, Benny, David Breitgand, Eliezer Levy, Alex Galis, Kenneth Nagin, Ignacio Martín Llorente, Rubén Montero, et al. 2009. The Reservoir Model and Architecture for Open Federated Cloud Computing. *IBM Journal of Research and Development* 53 (4): 4:1–4:11.

Roth, Alvin E., ed. 1988. *The Shapley Value: Essays in Honor of Lloyd S. Shapley*. Cambridge University Press.

Samaan, Nancy. 2014. A Novel Economic Sharing Model in a Federation of Selfish Cloud Providers. *IEEE Transactions on Parallel and Distributed Systems* 25 (1): 12–21.

Tak, Byung-Chul, Bhuvan Urgaonkar, and Anand Sivasubramaniam. 2011. *To Move or Not to Move: The Economics of Cloud Computing*. HotCloud.

Truong, Hong-Linh, and Schahram Dustdar. 2010. Composable Cost Estimation and Monitoring for Computational Applications in Cloud Computing Environments. *Procedia Computer Science* 1 (1): 2175–2184.

Wang, Wei, Baochun Li, and Ben Liang. 2012. *Towards Optimal Capacity Segmentation with Hybrid Cloud Pricing*. 2012 IEEE 32nd International Conference on Distributed Computing Systems, 425–434. IEEE.

CHAPTER 6

Value Creation and Power Asymmetries in Digital Ecosystems: A Study of a Cloud Gaming Provider

Arto Ojala, Nina Helander, and Pasi Tyrväinen

Abstract Digital platforms connecting users and service providers have a central role in determining the value creation structure of ecosystems. Platform developers try to achieve a dominant position for the platform with a strong ecosystem around it. The size and attractiveness of the services can attract new users, and growing user volume can bring new co-operative service providers to the service partner network. An interesting question is how the presence of power and potential power asymmetry affect the value

A. Ojala (✉)
School of Marketing and Communication, University of Vaasa, Vaasa, Finland
e-mail: arto.ojala@univaasa.fi

N. Helander
Faculty of Management and Business, Tampere University, Tampere, Finland
e-mail: nina.helander@tuni.fi

P. Tyrväinen
Faculty of Information Technology, University of Jyväskylä, Jyväskylä, Finland
e-mail: pasi.tyrvainen@jyu.fi

© The Author(s) 2020   89
T. Lynn et al. (eds.), *Measuring the Business Value of Cloud Computing*, Palgrave Studies in Digital Business & Enabling Technologies, https://doi.org/10.1007/978-3-030-43198-3_6

creation capability and the structure of a network around a platform? This chapter describes an example of value creation and the influence of power asymmetry in a digital ecosystem built around a cloud gaming platform.

Keywords Digital ecosystems • Digitalisation • Digital platforms • Partner networks

6.1   INTRODUCTION

The digitalisation of artifacts provides new opportunities that change traditional business models and how services are delivered to end-users (Baber et al. 2019; Ojala 2016; Yoo et al. 2010). In particular, digital platforms that enable emergence of new kinds of ecosystems are changing the way people interact with digital technology (Adner 2012; Yoo 2010). These ecosystems can be conceptualised as "a loosely coupled network of actors who interact and offer resources of different kinds, which together form a digital service around the platform" (Ojala et al. 2018, p. 729). In ecosystems, digital platforms have a central role that determines the structure of the ecosystem. Achieving a dominant position for a platform and building a strong ecosystem around the platform is a demanding process in which value creation has a central role.

Compared to the traditional value creation chains where value moves from a firm to customers (Porter 1985), digital platform providers must consider how value is generated for multiple sides of the platform (Eisenmann et al. 2006). Furthermore, actors on the different sides of the platform depend on the size and attractiveness of the other side of the market (Adner 2012). For instance, in the videogame industry where competition among various gaming platforms is intense (Lee 2012), game studios are more likely to develop games for platforms that have a lot of existing players. In line, video game players tend to favour gaming platforms that provide a high volume of interesting game content. Developing a platform that provides value to multisided markets and building a strong ecosystem is a complex and demanding process (Ojala and Lyytinen 2018). Even if a firm has an excellent innovation for a gaming platform, the firm's value creation still largely relies on other innovations within the ecosystem (Adner 2012; Lee 2012) and the power the firm has over other firms.

To better understand value creation and the concept of power in digital ecosystems, this chapter examines the value creation literature (Allee

2000) and management studies on organisational power (Astley and Sachdeva 1984; Mintzberg 1978). Specifically, we contribute to understanding of this topic in the context of digital platforms by studying (1) what kind of direct and indirect value is generated in the focal partner network, (2) how the focal network and the power positions evolve over time, and (3) how power asymmetry influences value creation within the network. We focus on the videogame industry because it has multisided markets, and ecosystems have a strong role (Lee 2012; McIntyre and Srinivasan 2017). Further, the industry has a relatively long history with well-established gaming platforms (Lee 2012). This makes the entry of newcomers challenging as they may have very little power in the market, and they have to create new ecosystems from scratch (Ojala 2016).

6.2 DIGITAL PLATFORMS AND ECOSYSTEMS

Digital platforms are generally organised in a network-like architecture, referred to as a layered modular structure with loosely coupled interfaces (Yoo et al. 2010). The architecture forms a hybrid of a modular structure and a layered structure (cf. Ulrich 1995). The architecture emerges when digital components and functions form the primary platform services or when the components and functions are embedded in hierarchically organised product structures (Yoo et al. 2010). In the loosely coupled, multi-layered architecture, the digital platform is organised in four layers: (1) device, (2) network, (3) service, and (4) content. The device layer refers to physical devices that connect and interact with the platform and its services, such as a television set, a mobile phone, or a gaming console. The network layer refers to the networking protocols that the platform offers to communicate over the networks to devices at the device layer. The service layer relates to the functionality of the applications that run on the platform and that enable users to use the content across different devices. The content layer covers the content that customers interact with, such as music, games, or videos. These layers form an ecosystem around the digital platform where several diverse actors, such as platform owners, content providers, telecom operators, device manufacturers, end-users, etc. (Koch and Windsperger 2017; Tiwana 2013), may participate, create value, and form multisided markets (Eisenmann et al. 2006). All these actors shape the competition and power asymmetries around the digital platform because each actor has unique interests and motivations for participating in the ecosystem.

6.3   Value Creation Within Digital Ecosystems

In this chapter, we use term "partner network" to refer to various actors within the ecosystem that cooperate directly with the focal firm. To operate successfully in the ecosystem, a firm must recognise the potential and current actors in the ecosystem that contains the firm's partner network (Ojala and Helander 2014). Thus, a partner network is a more focused part of the larger digital network. In the platform context, the network has the characteristics of the triangular structure typical of two-sided markets (Eisenmann et al. 2006). These partners can usually be grouped into producers and customers, and in the gaming industry, into game developers and consumers. Between these two main actor groups, the platform provider as the focal firm acts as an intermediary for creating value. Further, the focal firm needs to identify the value of their own offering, and how this value can be delivered to benefit other actors in the network. Thus, a firm should map all the actors in the ecosystem that could benefit from the firm's offering and the resources that the firm will need to commercialise their service (Allee 2000).

Created value can be defined as a trade-off between benefits and sacrifices (Lapierre 2000) that can be monetary or non-monetary. Monetary benefits and costs are usually easier to measure. However, the role of non-monetary rewards and costs in value perception is significant, too. Non-monetary rewards can be a status reward, emotional reward, or gain of new competences, while non-monetary costs may include the time, effort, energy, and amount of conflict customers engage in to obtain the product or service (Walter et al. 2001).

Value creation from a functional perspective (Walter et al. 2001) offers a view on the types of activities that actors may perform to create more value for network members. According to function-oriented value analysis, a firm may gain value from relationships with direct and indirect functions. Direct value can be produced through profit, volume, and safeguard functions. (Walter et al. 2001). For example, a safeguard function is a direct value creation function: If the firm has a long contract with the customer, this relationship creates safeguard value for the firm. Indirect functions, in contrast, require the input of third parties. Indirect functions include market, access, and innovation functions. For example, the market function means that one actor gives access to another market area with new potential partner actors.

6.4 THE CONCEPT OF POWER IN PARTNER NETWORKS

Astley and Sachdeva (1984) identified three sources of power: hierarchical authority, resource control, and network centrality. Hierarchical authority often relates to official positions that actors have over one another, so they are usually coupled with actors like authorities or supervisors (Astley and Sachdeva 1984). Resource control looks at the environment of an organisation, as it states that everyone is dependent on the resources of others: "organizations are open social systems that require a supply of resources from the environment in order to sustain their operations" (Astley and Sachdeva 1984, p. 106). Thus, no organisation or actor can act alone. Power based on resources is naturally higher in the case of critical or hard-to-obtain resources than in bulk resources. The third source of power, network centrality, refers to the position of an actor in a network (Easton 1992). Håkansson and Snehota (1989) argued that network actors aim to increase their own power and influence in networks as the actors believe that more powerful positions within a network will enable the actors to achieve other objectives. An actor's power and position within a network are closely related. In the end, power is realised in the interaction processes that form a relationship (Turnbull and Valla 1986); there cannot be power without the other part of the relationship.

6.5 RESEARCH METHOD

This chapter examines a qualitative, longitudinal case. A case study provides detailed (Edmondson and Mcmanus 2007) and empirically rich data (Eisenhardt and Graebner 2007) connected to a complex phenomenon. A longitudinal case study also facilitates examination of the evolution of a firm's activities (Eisenhardt 1989) and partner networks (Ford and Redwood 2005). In the data collection, we combined interview and secondary material covering the whole history of the firm from 2000 to 2015. We conducted 15 interviews, each lasting 45–90 minutes. Secondary data sources included the firm's brochures and press releases, which provided an extensive and detailed historical description of the firm. By using this information, we also triangulated the information. In the data analysis, we followed the steps: (1) data condensation, (2) data display, and (3) drawing and verifying conclusions as recommended by Miles and Huberman (1994).

6.6   G-Cluster

The case firm, G-cluster, develops games-on-demand services. Throughout the history, G-cluster has been small, employing 10–40 persons. Compared to traditional videogame platform providers (see Lee 2012), G-cluster's business idea is based on a completely different way of providing video-games to players. Traditionally, games are installed on a computer or a game console that runs the game. In the G-cluster business model, games are run on the platform operated by telecom operators on their game serv-ers. The game server transmits the game content to the end users' devices over the broadband network. The client devices receive the stream, display the game, and transmit users' commands back to the game server. Thus, G-cluster's gaming platform makes it possible to bring games to the cloud environment. The commercialisation has been challenging as the industry is dominated by large and well-known gaming platform providers making market entry for newcomers difficult. However, as G-cluster was the first cloud-based gaming platform provider in the market, the firm's innova-tion attracted increasing interest among potential partners.

6.6.1   *Creation of a Partner Network*

G-cluster's business goal was new in 2000 and took time to evolve, thus the first real partner network emerged in 2005 (Fig. 6.1). The figure shows the key partners within the ecosystem that G-cluster acted with directly (straight arrow) when the firm commercialised its service. First, G-cluster needed content for the gaming platform. To acquire game con-tent, which was critical for G-cluster's service, the firm partnered with game publishers and licensed games for the platform on a revenue-sharing basis.

Revenue sharing gave G-cluster access to a portfolio of games while protecting the firm's cash flow. However, the potential revenue was not, in many cases, appealing as a primary partnering factor for the game pub-lishers. As a result, G-cluster had to demonstrate other benefits—benefits that would bring value to the game publishers—if they provided games for the platform (see also Ojala and Tyrväinen 2011a, b). G-cluster motivated game studios by emphasising the benefits of cloud computing, like avoid-ing piracy, illegal copying, and second-hand markets for the games.

G-cluster also needed partners that are capable of running their cloud gaming service on their servers and provide access for players to services.

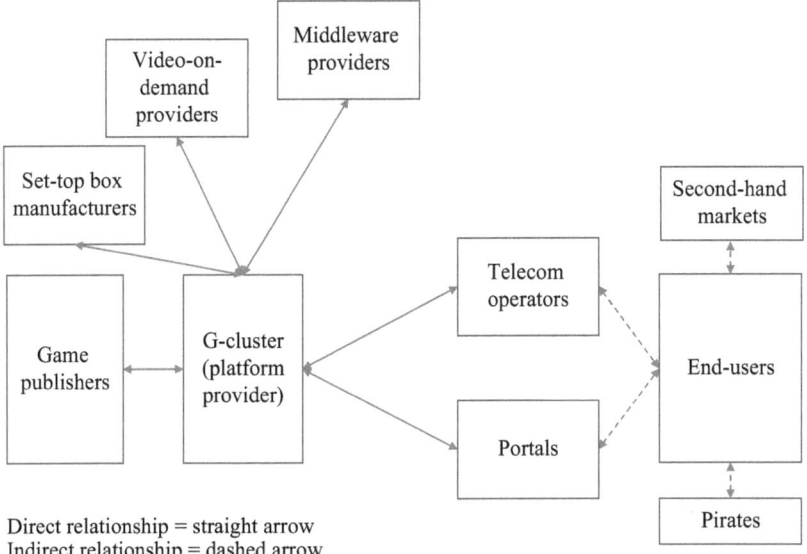

Fig. 6.1 The partner network in 2005

Consequently, G-cluster developed relationships with telecom operators that became important partners within the ecosystem. Telecom operators had good marketing channels and a large, existing customer base. As they are big players in the market, they also offered more visibility and a brand name that could be used for marketing purposes. Operators also motivated game studios to make their games available on G-cluster's platform. However, the move toward cooperation with telecom operators was not easy due to their powerful market position. G-cluster needed to demonstrate the value of their product for telecom operators. In addition to monetary benefits, G-cluster's service offered a good opportunity to extend the telecom operators' existing product portfolio and differentiate their offering from that of competitors.

To commercialise the service and to partner with telecom operators, G-cluster needed resources from video-on-demand service providers, set-top box manufacturers, and middleware software providers. For video-on-demand service providers, G-cluster's game platform offered new functionalities and enabled them to offer more content for telecom operators. Set-top box manufacturers needed new functions for their devices, and G-cluster's gaming platform brought extra value. Middleware software providers, which sell software to telecom operators, benefited from G-cluster's platform, as they were able to integrate game-on-demand services in the telecom operators' set-top boxes. The cooperation among these three different types of firms was based mainly on the mutual benefits that the partner network provided rather than on monetary benefits. These relationships were symmetric in the power aspect.

Portals (like Yahoo) were also important partners within the partner network that gave PC users access to G-cluster's cloud gaming service, and enabled multihoming of the service. The portals also took care of marketing activities and charged customers via the portals' invoicing systems. For the portals, G-cluster's cloud service generated revenue without requiring any investment, and the service was easily integrated with their current business.

6.6.2   *Evolution in Partner Networks*

Over a five-year period, changes in the ecosystem, markets and G-cluster's services impacted the partner network (Fig. 6.2). Due to the increasing competition in the PC game markets and the emergence of free-to-play

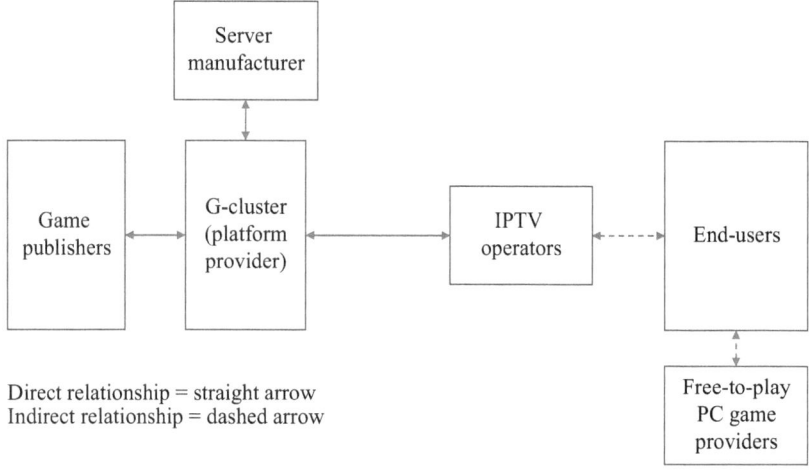

Fig. 6.2 The partner network in 2010

games, G-cluster focused solely on Internet Protocol Television (IPTV) networks and users and removed the PC market from the network. In addition, the operators had increased their IPTV offering remarkably, and there was ongoing growth in the market as IPTV connections became more reliable.

G-cluster also developed its product further, making it a ready-made service for operators. For instance, G-cluster developed their own billing system and a user interface (menu) that players used to select games from the G-cluster virtual games store. By including these features, G-cluster became less dependent on other firms and strengthened their position in the partner network.

In 2010, G-cluster established a relationship with a large and well-known server manufacturer that provided cloud technologies for telecom operators. This relationship provided mutual benefits for both firms. The main point of this partnership was mutual value. For G-cluster, the cooperation brought in a more reliable and influential partner, one that could market the gaming service further to telecom operators. This reliable partner increased G-cluster's marketing and sale resources considerably. Conversely, by including G-cluster's cloud gaming feature on the server manufacturer's offering, they obtained added value that the manufacturer was able to use when they marketed their servers to operators.

6.6.3   Equal Power in a Partner Network

To multihome the service beyond IPTV operators, G-cluster brought their own cloud gaming console to the market in 2015. The console was a small physical device that connected G-cluster's game service to players' TVs without an IPTV subscription. This console made G-cluster's gaming service available through any telecom operator, as the service was no longer tied to IPTV. Another change in the partner network was that G-cluster no longer cooperated with the server manufacturer (Fig. 6.3). In 2015, G-cluster had well-known operators as reference customers and no longer needed the server manufacturer or other third parties to reassure operators about the service. Thus, although the number of actors in the partner network decreased, G-cluster increased their number of critical partners. This change led to a more stable and co-operative network, where power is more equally dispersed among the actors.

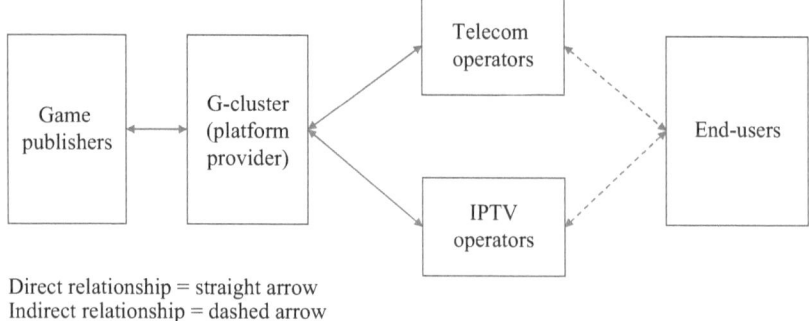

Direct relationship = straight arrow
Indirect relationship = dashed arrow

Fig. 6.3   The partner network in 2015

6.7   RESULTS

The network's ongoing evolution was evident in the case study. This process is typical for networks, caused by a change in the ecosystem, value creation logic, and the consolidation of relationships between actors. If a firm is to gain a better position in a partner network and market the offering, the firm must cooperate with several kinds of partners. The firm must demonstrate the value of the offering to several partners in the network. This evolution in the network position also affects the power setting. The sources of power and the direct and indirect values between the various actors in each micro-position are shown in Table 6.1.

The direct value was related, in addition to monetary value, to the critical resources that G-cluster needed to commercialise its service. The direct value can be divided into resources enabling the service (provided by game studios and telecom operators) and functionalities needed for the product. For the partners that enabled G-cluster's service, G-cluster provided mainly financial benefits as a direct value. Although indirect value was not critical to G-cluster's service, this value provided substantial help in marketing and networking. Indirect value worked similarly: G-cluster gained resources for marketing and networking, and its partners gained a new feature for their services, one that the partners were able to use in their marketing.

Table 6.1 Direct and indirect value and the element of power in each focal relationship (extended from Ojala and Helander 2014)

Focal relationship	Direct value	Indirect value	Relationship power and position	Source of power
G-cluster → game publishers	• Financial value • SDK	• No piracy • No second-hand markets • Bigger markets for games • New revenue models	• Game publisher had a strong position in 2005. • Later, the relationship was more equal as G-cluster reached the position of critical solution supplier.	• Game publishers provided content for G-cluster's platform. • G-cluster provided critical competences in developing markets.
Game publishers → G-cluster	• Content (games) for the platform	• References		
G-cluster → telecom/IPTV operators	• Financial value • More content for the services	• Market potential	• Telecom/IPTV operators had a strong position that remains strong in the changing situation.	• Telecom/IPTV operators operated G-cluster's platform.
Telecom/IPTV operators → G-cluster	• Delivery channel • Computing services	• Pre-existing customers • Marketing • Brand name • Networking with game publishers		• Telecom/IPTV operators had many customers for the service and the marketing channel.

(continued)

Table 6.1 (continued)

Focal relationship	Direct value	Indirect value	Relationship power and position	Source of power
G-cluster → video-on-demand service providers Video-on-demand service providers → G-cluster	• New feature for the service • Invoicing system for the service	• Market potential • Networking with IPTV providers	• Equal	• Video-on-demand service providers had contacts with telecom/IPTV operators. • G-cluster offered new features that made video-on-demand providers' products more attractive to telecom/IPTV operators.
G-cluster → set-top box manufacturers Set-top box manufacturers → G-cluster	• New feature for the service	• Market potential • Networking with IPTV providers	• Equal	• Set-top box manufacturers had contacts with telecom/IPTV operators. • G-cluster offered new features that made set-top box manufacturers' products more attractive to telecom/IPTV operators.

Relationship	Value created	Power source	Power balance	Outcome
G-cluster → middleware software providers Middleware software providers → G-cluster	• New feature for the service	• Market potential • Networking with IPTV providers	• Equal	• Middleware software providers had contacts with telecom/IPTV operators. • G-cluster offered new features that made middleware software providers' services more attractive to telecom/IPTV operators.
G-cluster → portals Portals → G-cluster	• Financial value • More content for the services • Delivery channel • Games menu	• Market potential • Pre-existing customers • Marketing • Brand name	• Equal	• Portals got more content for their services. • G-cluster got a delivery channel for their service.
G-cluster → server manufacturer Server manufacturer → G-cluster	• New feature for the services	• Market potential • Networking with telecom/IPTV operators	• Equal	• Server manufacturer had contacts with telecom/IPTV operators. • G-cluster offered new features that increased the value of the server manufacturer's product.

As can be observed from Figs. 6.1, 6.2, 6.3, the partner network evolved from a complex structure involving several actors (in 2005) to a simple network structure that included only the most critical partners (in 2015). Easton (1992) observed that the more the independence of a firm increases, the less fragmented its network becomes. In the G-cluster case, the number of partners in the network decreased over time, but the number of content and telecom operators increased. The firm's network evolved through the following steps. First, the platform provider networked with all possible actors operating within the different layers of the multilayered platform architecture (Yoo et al. 2010) that could benefit from the platform to achieve market visibility. Second, after getting more visibility in the ecosystem, the firm focused on the most powerful actors within the content layer (game studios) and network layer (telecom operators), and developed deeper relationships with them. These actions increased the platform provider's position in the partner network and made it more concentrated.

Power is increased when the focal actor has more alternatives to choose from, and the power position relates to the size of the firm. In the beginning, the telecom operators and game publishers were big companies that had many content and current technology suppliers from which to choose. However, as the technological landscape changed to favour G-cluster's solution and competences, G-cluster achieved a stronger position in the network and became the critical supplier. Critical competences and technological changes in the market acted as the trigger for changes in the network structure and the power positions. Thus, a link between value creation capability and the power position was visible through G-cluster's competences. The direct financial value and the indirect value that G-cluster provides to game publishers and telecom operators exceeded the competitors' value creation capability and led to closer co-operation.

6.8   Conclusion

This chapter contributes understanding of digital ecosystems in several ways. First, the chapter incorporated network theory (Johanson and Mattsson 1988), management studies related to organisational power

(Astley and Sachdeva 1984; Mintzberg 1978), and literature on value creation (Walter et al. 2001). The chapter conceptualised the value creation and evolution of the partner network in the contemporary context of digital ecosystems. Specifically, the chapter provides an in-depth view of how a platform provider can create value with other actors in the network and how the value can be used to form a good network position and power symmetry in the market.

Second, this chapter provides detailed insights into how and why a network changed over 10 years. This is important in developing a realistic view of value creation (Walter et al. 2001) and network development. As Halinen and Törnroos (2005) noted, over time networks change in relation to the problems that they aim to solve and, in this way, to the value they aim to create. When situations change, new kinds of actors may be needed in network cooperation, and this will lead to a change in the network structure.

Third, this chapter provides detailed knowledge on the range of direct and indirect value that network actors create for each other and finally, for the end customer. In addition, the chapter sheds light on the interesting relation between value creation capability and power. Monetary and non-monetary value creation capabilities are needed to ensure the development of a stronger network position and to balance the asymmetric power setting that the larger firm may have in the relationship. Thus, through critical competence development that leads to increased value creation capability, even a small actor may enhance its network position.

REFERENCES

Adner, Ron. 2012. *The Wide Lens: What Successful Innovators See That Others Miss.* Penguin.

Allee, Verna. 2000. Reconfiguring the Value Network. *Journal of Business Strategy* 21 (4): 36–39. https://doi.org/10.1108/eb040103.

Astley, W. Graham, and Paramjit S. Sachdeva. 1984. Structural Sources of Intraorganizational: Power: A Theoretical Synthesis. *Academy of Management Review* 9 (1): 104–113.

Baber, William W., Arto Ojala, and Ricardo Martínez. 2019. *Transition to Digital Distribution Platforms and Business Model Evolution.* 52nd Hawaii International Conference on System Sciences. http://scholarspace.manoa.hawaii.edu/handle/10125/59937.

Easton, Geoffrey. 1992. Industrial Networks: A Review. In *Industrial Networks: A New View of Reality,* 3–27. London and New York: Routledge.

Edmondson, Amy C., and Stacy E. Mcmanus. 2007. Methodological Fit in Management Field Research. *Academy of Management Review* 32 (4): 1155–1179. https://doi.org/10.5465/AMR.2007.26586086.

Eisenhardt, K.M. 1989. Building Theories from Case Study Research. *Academy of Management Review* 14 (4): 532–550.

Eisenhardt, K.M., and M.E. Graebner. 2007. Theory Building from Cases: Opportunities and Challenges. *Academy of Management Journal* 50 (1): 25.

Eisenmann, Thomas, Geoffrey Parker, and Marshall W. Van Alstyne. 2006. Strategies for Two-Sided Markets. *Harvard Business Review* 84 (10): 92–101.

Ford, David, and Michael Redwood. 2005. Making Sense of Network Dynamics Through Network Pictures: A Longitudinal Case Study. *Industrial Marketing Management* 34 (7): 648–657. https://doi.org/10.1016/j.indmarman.2005.05.008.

Håkansson, Håkan, and Ivan Snehota. 1989. No Business is an Island: The Network Concept of Business Strategy. *Scandinavian Journal of Management* 5 (3): 187–200. https://doi.org/10.1016/0956-5221(89)90026-2.

Halinen, Aino, and Jan-Åke Törnroos. 2005. Using Case Methods in the Study of Contemporary Business Networks. *Journal of Business Research* 58 (9): 1285–1297.

Johanson, J., and L.G. Mattsson. 1988. Internationalisation in Industrial Systems: A Network Approach. In *Strategies in Global Competition,* 287–314. London: Croom Helm.

Koch, Thorsten, and Josef Windsperger. 2017. Seeing Through the Network: Competitive Advantage in the Digital Economy. *Journal of Organization Design* 6 (1): 6.

Lapierre, Jozée. 2000. 'Customer-Perceived Value in Industrial Contexts. *Journal of Business & Industrial Marketing* 15 (2/3): 122–145. https://doi.org/10.1108/08858620010316831.

Lee, Robin S. 2012. Home Videogame Platforms. In *The Oxford Handbook of the Digital Economy,* 83–107. Oxford: Oxford University Press.

McIntyre, David P., and Arati Srinivasan. 2017. Networks, Platforms, and Strategy: Emerging Views and next Steps. *Strategic Management Journal* 38 (1): 141–160. https://doi.org/10.1002/smj.2596.

Miles, Matthew B., and A. Michael Huberman. 1994. *Qualitative Data Analysis: An Expanded Sourcebook.* SAGE.

Mintzberg, Henry. 1978. Patterns in Strategy Formation. *Management Science* 24 (9): 934–948.

Ojala, Arto. 2016. Business Models and Opportunity Creation: How IT Entrepreneurs Create and Develop Business Models under Uncertainty. *Information Systems Journal* 26 (5): 451–476. https://doi.org/10.1111/isj.12078.

Ojala, Arto, and Nina Helander. 2014. *Value Creation and Evolution of a Value Network: A Longitudinal Case Study on a Platform-as-a-Service Provider.* 2014 47th Hawaii International Conference on System Sciences, 975–984. https://doi.org/10.1109/HICSS.2014.128.

Ojala, Arto, and Kalle Lyytinen. 2018. *Competition Logics During Digital Platform Evolution.* 51st Hawaii International Conference on System Sciences. http://scholarspace.manoa.hawaii.edu/handle/10125/50017.

Ojala, Arto, and Pasi Tyrväinen. 2011a. Value Networks in Cloud Computing. *Journal of Business Strategy* 32 (6): 40–49. https://doi.org/10.1108/02756661111180122.

———. 2011b. Developing Cloud Business Models: A Case Study on Cloud Gaming. *IEEE Software* 28 (4): 42–47. https://doi.org/10.1109/MS.2011.51.

Ojala, Arto, Natasha Evers, and Alex Rialp. 2018. Extending the International New Venture Phenomenon to Digital Platform Providers: A Longitudinal Case Study. *Journal of World Business* 53 (5): 725–739. https://doi.org/10.1016/j.jwb.2018.05.001.

Porter, Michael E. 1985. *Competitive Advantage: Creating and Sustaining Superior Performance.* New York: Free Press.

Tiwana, Amrit. 2013. *Platform Ecosystems: Aligning Architecture, Governance, and Strategy.* Newnes.

Turnbull, Peter W., and Jean-Paul Valla. 1986. Strategic Planning in Industrial Marketing: An Interaction Approach. *European Journal of Marketing* 20 (7): 5–20. https://doi.org/10.1108/EUM0000000004652.

Ulrich, Karl. 1995. The Role of Product Architecture in the Manufacturing Firm. *Research Policy* 24 (3): 419–440. https://doi.org/10.1016/0048-7333(94)00775-3.

Walter, Achim, Thomas Ritter, and Hans Gemünden. 2001. Value Creation in Buyer–Seller Relationships: Theoretical Considerations and Empirical Results from a Supplier's Perspective. *Industrial Marketing Management* 30 (4): 365–377. https://doi.org/10.1016/S0019-8501(01)00156-0.

Yoo, Youngjin. 2010. Computing in Everyday Life: A Call for Research on Experiential Computing. *MIS Quarterly* 34 (2): 213–231.

Yoo, Youngjin, Ola Henfridsson, and Kalle Lyytinen. 2010. The New Organizing Logic of Digital Innovation: An Agenda for Information Systems Research. *Information Systems Research* 21 (4): 724–735. https://doi.org/10.1287/isre.1100.0322.

Measuring the Business Value of Cloud Computing: Emerging Paradigms and Future Directions for Research

Theo Lynn, Pierangelo Rosati, and Grace Fox

Abstract Much of the research on measuring the business value of cloud computing examines cloud computing from the perspective of a central-ised commodity-based aggregated conceptualisation of cloud computing, largely based on the NIST reference architecture. Advances in new proces-sor architectures and virtualisation combined with the rise of the Internet of Things are not only changing cloud computing but introducing new computing paradigms from the cloud to the edge. These new paradigms present both opportunities and challenges, not least managing complexity several orders of magnitude greater than today. Yet, academic research on measuring the business value of cloud computing is lagging practice and remains far removed from these innovations. New research is required that explores the relationship between investments in new cloud computing

T. Lynn (✉) • P. Rosati • G. Fox
Irish Institute of Digital Business, DCU Business School, Dublin, Ireland
e-mail: theo.lynn@dcu.ie

© The Author(s) 2020   107
T. Lynn et al. (eds.), *Measuring the Business Value of Cloud Computing*, Palgrave Studies in Digital Business & Enabling Technologies, https://doi.org/10.1007/978-3-030-43198-3_7

paradigms and business value, and the measurement thereof. This chapter explores a selection of these new paradigms, which may provide fruitful research pathways in the future.

Keywords Function-as-a-Service • Serverless computing • Containerisation • High performance computing

7.1   INTRODUCTION

Our interest in measuring the business value of cloud computing stemmed from a 2017 multi-disciplinary survey of the literature we conducted (Rosati et al. 2017). Our findings at that time highlighted a number of worrying issues in the 53 papers published from 2009 to 2016 that we examined. Firstly, the overwhelming majority of studies, in both information systems (IS) and computer science (CS), focussed primarily on one service model, Infrastructure-as-a-Service (IaaS). This is unsurprising as it allows an easier comparison with traditional on-premise computing. In IS, there was also a tendency to conflate all service models as "the cloud" thereby missing on important nuances about how discrete service models and delivery models can deliver different types of business value. Secondly, there were significant differences between IS and CS papers with regards to the granularity and substantiation of impact of the IT artefacts studied. While CS papers examined IT artefacts at an extremely low level of granularity in a cloud solution stack when compared to IS papers, they could clearly link these artefacts across the causal chain to economic factors in a way that IS papers could not or did not. Furthermore, the impact was measurable in much shorter time horizons. Thirdly, the techniques used to measure business value were concentrated on measuring costs e.g. Total Cost of Ownership (TCO). This is not wholly unsurprising given that the focus was mostly IaaS and migration from on-premise. However, more concerning was that many of the studies, and, in particular, CS studies, demonstrated significant methodological issues in their calculation of costs, and where examined, benefits. In particular, few attempts were identified to measure intangible benefits.

In summary, our feeling in 2017 was that there was a need for a more systematic and interdisciplinary approach to researching the conceptualisation and measurement of the business value of cloud computing (BVCC) in a more disaggregated way. Assuming an unchanging technological landscape, this would have still been a major challenge. However, our conceptualisation of the "cloud" is radically changing. The pace of change in cloud computing and how enterprises manage and use it has accelerated dramatically in recent years. As a consequence, it is surely worth considering whether the nature of the business value created by cloud computing and how we measure it has changed too. This chapter presents a number of new paradigms in cloud computing, changes in cloud architectures, and research pathways we believe may prove promising avenues for future research for both IS and CS researchers.

7.2   THE CHANGING NATURE OF THE CLOUD

The accepted definition of cloud computing has not changed. It is:

> ...a model for enabling ubiquitous, convenient, on-demand network access to a shared pool of configurable computing resources (e.g., networks, servers, storage, applications, and services) that can be rapidly provisioned and released with minimal management effort or service provider interaction. (Mell and Grance 2011, p. 2)

However, the nature of the cloud has changed. It is increasing more abstracted, heterogeneous, composable, and automated.

7.2.1   The Evolution of Shared Resources

How resources are shared in the cloud is evolving rapidly (see Fig. 7.1). In the first phase of cloud computing, we saw a shift from monolithic architectures to service oriented architectures; this is what is largely described in the NIST Cloud Computing Reference Architecture (Mell and Grance 2011) and the focus of BVCC research from 2009 to 2016. In this phase, cloud service providers and their customers benefit from their own discrete virtual machines (VMs) running on shared infrastructure.

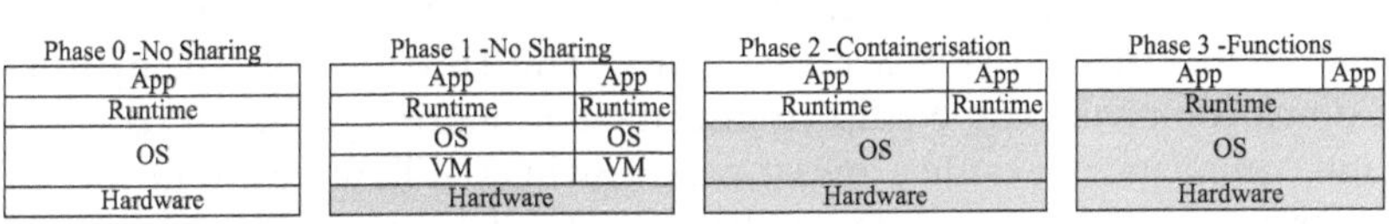

Fig. 7.1 Evolution of shared resources in cloud computing. Grey areas are shared. (Adapted from Hendrickson et al. 2016)

Since the open sourcing of dotCloud's container technology in 2013, the nature of the cloud began changing again. Containerisation enabled operating system (OS)-level virtualization where containers hold all the components necessary to run a specific software program and a minimal subset of an OS. As a concept, this results in a number of benefits relevant to measuring business value. For example, containers are less resource-intensive than VMs and therefore result in reduced operational expenditure. They are more portable thus reducing lock-in and increasing agility and flexibility. New services can be provisioned faster thus resulting in increased time to market. Despite these advantages, there is scant discussion of containerisation (or micro-services) in the IS literature and even less, if any, on the measurement of the business value of this architectural style.

More recently, we have seen the emergence of serverless cloud computing. Here, effectively all resources are pooled including hardware, operating systems and runtime environments. Serverless computing is "a software architecture where an application is decomposed into 'triggers' (events) and 'actions' (functions), and there is a platform that provides a seamless hosting and execution environment" (Glikson et al. 2017, p. 1). The software owner does not necessarily have to concern themselves with management of the runtime environment instead can focus on developing and deploying relatively lightweight, single purpose stateless functions that can be executed on-demand, typically through an API, without consuming any resources until the point of execution (Lynn et al. 2017). As such, this cloud service model is often called Function as a Service (FaaS). The cloud service provider assumes responsibility for data centre management, server management and the runtime environment. The software operators only pay for resources when they are executed thus reducing the cost of deployment dramatically. Furthermore, FaaS also transforms the business model of cloud service providers e.g. pricing at the level of execution runtime for computer code rather than how long an instance is running (Eivy 2017). For these reasons, FaaS is gaining significant traction. It has been adopted not only by the major hyperscale cloud service providers (e.g. Google, Microsoft, Amazon Web Services (AWS), and IBM) but also many well-known companies e.g. Netflix (transcoding, monitoring, disaster recovery, and compliance), Seattle Times (image resizing), Zillow (real-time mobile metrics), and Major League Baseball Advanced Media (data analysis, and player and game metrics) (Lynn et al. 2017). As previously mentioned, there was an existing need for more BVCC research relating to traditional cloud computing service models i.e. IaaS, PaaS, and SaaS (and to a much lesser extent Business Process as a Service—BPaaS); there is virtually no research on measuring the business value of containerisation (microservices) or serverless cloud computing (FaaS).

7.2.2   The Heterogeneous Cloud

As cloud computing continues to become the dominant computing paradigm, cloud service providers are looking for new segments for growth and enterprises are looking for new ways to create business value from migrating to the cloud. Two segments which have garnered a lot of attention in recent years are Big Data analytics and high performance computing (HPC) in the cloud. The benefits of Big Data and related analytics include increased agility (Ashrafi et al. 2019), innovation (Lehrer et al. 2018), and competitive performance (Côrte-Real et al. 2017; Mikalef et al. 2019) and is widely discussed in both IS and CS literature, and even more so in practice. The contribution of HPC is less widely discussed yet is recognised as playing a pivotal role in both science discovery and national competitiveness (Ezell and Atkinson 2016). The widespread use of both Big Data analytics and HPC have been hampered by significant upfront investment and indirect operational expenditure (including specialised staff) associated with running and maintaining these infrastructures. Big Data analytics and HPC in the cloud represent massive opportunities to unleash business value through reduced CapEx and OpEx as well as democratising Big Data and HPC infrastructure and tools and thus increase innovation output.

Traditionally, and to a large extent today, cloud computing systems are optimised to cater for multiple tenants and a large number of small workloads. The primary focus of traditional cloud computing is rapid scalability and as such is designed for perfectly or pleasingly parallel problems (Lynn 2018). For such workloads, while servers must be available and operational, neither the precise physical server nor the speed of the connections between processors that executes a request is important provided the resource database remains coherent (Eijkhout et al. 2016). Unfortunately, for Big Data analytics or HPC workloads, enterprise users typically require servers to be available on-demand and connected via high-speed, high-throughput, and low-latency network interconnects (Lynn 2018).

Heterogeneous computing refers to architectures that allow the use of different hardware types to work efficiently and cooperatively together.

Unlike traditional cloud infrastructure built on the same processor architecture, heterogeneity assumes use of different or dissimilar processors or cores that incorporate specialised processing capabilities to handle specific tasks more faster and more energy efficiently than general purpose processors (Scogland et al. 2014). For example, field-programmable gate arrays (FPGAs) and graphics processing units (GPUs) are co-processor architectures with relatively positive computation/power consumption ratios that offer significant performance and energy efficiency gains for Big Data analytics and HPC respectively. Increasingly heterogeneous computing is being extended beyond different processor architectures to include different networking infrastructure that can support higher throughput and lower latency (Shan 2006; Yeo and Lee 2011). In recent years, major public cloud providers including AWS, Microsoft Azure and Google Cloud offer specialist cloud services for Big Data and HPC uses cases built on heterogeneous clouds. These specialist clouds are increasingly being adopted by some of the world's largest companies including Aon (financial simulation), AstraZeneca (genome processing), BP (linear programming models), Disney (video streaming analytics and rendering), and Volkswagen (computation fluid dynamics). Despite the increasing availability and use of heterogeneous cloud computing, there is little research on the business value of adopting heterogeneous cloud computing.

7.2.3   The Composable Cloud

As more and more enterprises embrace digital transformation, even when private clouds are adopted, traditional IT architectures struggle to accommodate the cloud computing requirements from next generation applications. Legacy applications require infrastructure resiliency and exploit virtualization and clustering for portability and application state preservation. In contrast, next generation applications (NGAs) are designed to be horizontally scalable, containerised and continuously updated (Nadkarni 2017). IDC suggest that in most enterprise data centres, infrastructure is 45% over provisioned, 45% utilized, and 40% compliant with stated service level agreements (Nadkarni 2017).

Composable architectures assume that resources (e.g. compute, memory, storage, networking etc.) can be decoupled from the hardware they reside on and assembled and re-assembled using a control software layer to meet exact workload requirements on-demand (Ferreira et al. 2019). Once hardware is no longer required it can be released for use for another workload. There are a number of advantages to this approach. Firstly, discrete servers do not need to be configured for a specific application but rather hardware resources can be pooled to meet both legacy and NGAs dynamically. If more resources are needed to deliver a given workload, it is automatically provisioned. Secondly, composable architectures support heterogeneous computing and pools these resources in the same way thereby allowing enterprises to exploit the performance or energy efficiencies of these specialist resources. Thirdly, as each workload is provisioned exactly as needed, over-provisioning is reduced dramatically hereby reducing both CapEx and OpEx.

The Composable Cloud is a fundamentally different way to operate data centres and private clouds. Given that it reduces overprovisioning and related inefficiencies dramatically as well as freeing up valuable enterprise resources, not least cash flow and staffing, it is worthy of investigation by business value researchers.

7.2.4 The Automated Cloud

A side effect of new service models, increased heterogeneity, and composability is greater complexity in terms of reliability, maintenance and security (Marinescu 2017). This is particularly the case for large-scale enterprise systems and hyperscale cloud services where the scale of infrastructure, applications, and number of end users is significant. It is no longer feasible for IT teams to cost-effectively foresee and manage manually all possible configurations, component interactions, and end-user operations on a detailed level due to high levels of dynamism in the system (Lynn 2018). As such, enterprise IT and cloud service providers are increasingly looking to machine learning and artificial intelligence (AI) to manage this complexity but also automate previously manual tasks, and free up staff.

AIOps (AI for IT Operations) uses algorithms and machine learning to dramatically improve the monitoring, operation, and maintenance of distributed systems (Cordoso 2019). The main use cases for AIOps are performance analysis, anomaly detection, event correlation and analysis, and IT service management and automation with the ultimate goals of ensuring high service quality and customer satisfaction, boosting engineering productivity, and reducing operational costs (Prasad and Rich 2018; Dang et al. 2019). IDC's Worldwide Developer and DevOps 2019 Predictions suggest that by 2024, 60% of firms will have adopted AIOps (Gillen et al. 2018). Much of the market demand for AIOps is couched in the fear of outages and the ability of machine learning to predict such outages and enable preventative action to be taken before customers or business is impacted. Yet despite this optimism, there are significant challenges with the adoption of AIOps including changes in innovation methodologies including understanding business value and constraints, engineering mindsets, and engineering practices including data quality (Dang et al. 2019). From a business value research perspective, machine learning and AI poses additional challenges as the black box nature of these technologies can make interrogation and interpretability difficult.

7.3   Cloud Computing and the Internet of Things

Over the last five years, interest in the Internet of Things (IoT) has increased dramatically, partially fuelled by the increasing ubiquity of internet access and smartphones but also estimates of the value of IoT forecast to exceed $19 trillion over time (Cisco 2013a, b). This value is generated through connecting a fraction of the 1.4 trillion things *in situ* today, and consequently improving asset utilization, employee productivity, supply chain and logistics, customer experience, as well as accelerating innovation (Lynn et al. 2018).

Haller et al. (2009) define IoT as:

> A world where physical objects are seamlessly integrated into the information network, and where the physical objects can become active participants in business processes. Services are available to interact with these "smart objects" over the Internet, query their state and any information associated with them, taking into account security and privacy issues. (Haller et al. 2009, p. 15)

Smart objects may range from sensors, with little storage and data processing power, to modern smartphones. IoT assumes that smart things can carry out, with minimal latency, some degree of data processing and collaborate with other devices and systems, some local and some remotely. As such, it assumes a continuum of computing activity from the cloud to thing (C2T) where computing resources can be located in the cloud, at the thing (edge computing), or somewhere in between (fog computing). As such, IoT effectively extends cloud computing from a centralised service architecture to a decentralised one. Table 7.1 below summarises key definitions of new computing paradigms along the C2T continuum.

Table 7.1 Definitions of edge, fog and mist computing. (Adapted from Iorga et al. 2018)

Concept	Definition
Edge computing	Edge computing is the network layer encompassing the end devices and their users, to provide, for example, local computing capability on a sensor, metering or some other devices that are network-accessible.
Fog computing	Fog computing is a layered model for enabling ubiquitous access to a shared continuum of scalable computing resources. The model facilitates the deployment of distributed, latency-aware applications and services, and consists of fog nodes (physical or virtual), residing between smart end-devices and centralized (cloud) services.
Mist computing	Mist computing is a lightweight and rudimentary form of fog computing that resides directly within the network fabric at the edge of the network fabric, bringing the fog computing layer closer to the smart end-devices. Mist computing uses microcomputers and microcontrollers to feed into fog computing nodes and potentially onward towards the centralized (cloud) computing services.

For enterprises, cloud service providers and cloud carriers (e.g. Tier 1 network operators), IoT introduces complexity at yet another higher order of magnitude. To meet the Quality of Service (QoS) and Quality of Experience (QoE) requirements of SLAs with customers and/or end users, service providers and cloud carriers need to decide where best to locate compute and storage resources along the C2T continuum. As such, enterprises need to consider the geographic distribution and mobility of smart objects and latency at each location, the heterogeneity of smart objects, interoperability and federation, the necessity and capability of real-time interaction, and the scalability and agility of federated fog-node clusters (Iorga et al. 2018).

Haller et al. (2009) suggest that there are two main sources where enterprises can derive business value from the IoT—real world visibility and business process decomposition. Firstly, they argue that the use of automated identification and data collection technologies will give enterprises unparalleled insights in to what is happening in the real world thus enabling high resolution management and the potential deeper and better business insights, more effective optimisation, and better decision making. Secondly, they argue that IoT combined with real world visibility allows the decomposition of business processes in to process steps (and associated computing resources) which can be distributed from the cloud to edge thus enabling the decentralisation of business processes resulting in increased scalability and performance, better decision making, and innovation. From a cloud computing perspective, IoT involves key technical decisions that can impact the business value generated for the enterprise e.g. how much infrastructure should be placed at different points across the C2T continuum? What applications (or if distributed, what application components) should be operated at the edge and which should not? How do these placement decisions impact business value?

7.4   Towards an Agenda for Business Value in Cloud Computing Research

Cloud computing is a key enabling technology. As can be seen above, its interaction with other technologies, for example machine learning and AI, mobile computing, IoT, and HPC accelerates innovation and as a result potential business value. Recently, a number of authors have suggested research gaps and questions that can guide future research on business value in cloud computing. In particular, Schryen (2013) calls for greater research to close three IS business value research gaps which are relevant to business value in cloud computing i.e. ambiguity and fuzziness of the 'IS business value' construct, neglected disaggregation of IS investments, and the IS business value creation process as a grey box. Paraphrasing Schryen, a number of research pathways for business value in cloud computing arise:

> *How can we yield a comprehensive, consistent and precise understanding of the multifaceted construct 'cloud computing business value'? How can the assessment of (internal and competitive) business value account for the context of evaluation, and in particular the firm, industry, and country environment and preferences of evaluators?* (Schryen 2013, pp. 151–152)

Schryen (2013) suggests 'IS business value' is ambiguous and fuzzy. As previously discussed, our own experience is that not only is IS business value ambiguous and fuzzy but the techniques for measuring business value are often ambiguous and, where documented, are not applied consistently or comprehensively in such a way to allow comparison. In addition, a more nuanced approach to defining the firm and industry is needed. At a basic level, firms may include enterprises adopting cloud computing, cloud service providers, cloud carriers (e.g. network operators), cloud brokers, cloud auditors, and indeed edge device consumers (See Fig. 7.2 below). Each of these actors may operate in different industries and thus provide different industry and country contexts and associated constraints, particularly with respect to operation. For example, cloud computing is, by and large, a cross-border phenomenon however national data privacy laws, amongst others, create opportunities and risks for business value generation and capture.

How can total cloud computing investments be disaggregated conceptually and empirically such that the impact of different types of investments on the economic performance of the firm can be determined? How can the disaggregation of total cloud computing investments account for synergies and complementarities? (adapted from Schryen 2013, pp. 153–154)

This assumes one can disaggregate cloud computing investments from wider IS investments and then specific cloud computing investments. As indicated earlier, at a basic level this could be by service model (IaaS, PaaS, SaaS, and FaaS) or even by deployment model (private, public, hybrid, and community clouds), components of the extended cloud computing conceptual reference model (see Fig. 7.2), or a combination of all of these.

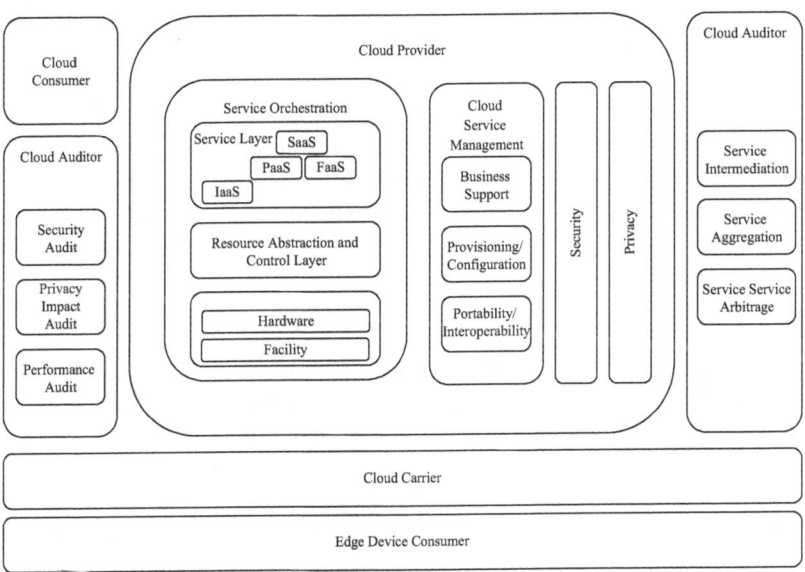

Fig. 7.2 Extended cloud computing conceptual reference model. (Adapted and extended from Liu et al. 2011)

To address these research questions, it assumes (1) a sufficiently detailed taxonomy of cloud computing investments can be catalogued, (2) critical success factors (CSFs) and key performance indicators (KPIs) can be mapped to these supporting assets, and (3) occurrences of synergies between different types of assets can be identified (Schryen 2013). This may require examination at a lower level of granularity than IS researchers typically undertake and as such may require CS support thus mandating interdisciplinary research.

How, why and when do cloud computing assets, cloud computing capabilities, IS assets and capabilities, and socio-organisational capabilities affect each other and joint create internal value? How, why, and when do cloud computing assets, cloud computing capabilities, IS assets and capabilities, and socio-organisational capabilities create competitive value, thus performing a value creation process? (adapted from Schryen 2013, p. 156)

This research question recognises that cloud computing assets and capabilities are a subset of wider IS assets and capabilities and have a bidirectional relationship with socio-organisational capabilities. This is particularly the case when we consider emerging use cases including IoT, Big Data analytics and HPC. It also recognises that value is created over time and that some aspects are static and some are dynamic. In the context of cloud computing, firms more than likely inherit the assets and capabilities of the chain of service provision and thus for a given time, have compound capabilities or what Carroll et al. (2013, 2014) call a composite capability. The business value of such capabilities is dependent on a number of socio-organisational factors, not least size, which obviously changes over time. As such, research must consider a contingency approach to business value.

7.5   CONCLUDING REMARKS

This chapter presents a number of new paradigms in cloud computing, changes in cloud architectures, and research pathways in business value in cloud computing research that we believe may provide future avenues of research for both IS and CS researchers. This is by no means exhaustive. Indeed, other chapters in this book cover aspects of business value in cloud computing research that could provide a fruitful stream of research. As we develop our understanding of cloud computing and the dependencies between cloud computing and other technologies (not least mobile, Big

Data, and IoT) the need for greater clarity on the definition and appropriate metrics of business value; robust business value measurement techniques; disaggregation of IS assets (and specifically cloud infrastructure); and the relationship between cloud assets and capabilities, other IS assets and capabilities, and socio-organisation capabilities, is required. This will require a deep understanding of these technologies and most likely collaboration between information systems and computer science researchers. More importantly, it will require a change in the mindsets of business value researchers in both disciplines.

References

Ashrafi, A., A.Z. Ravasan, P. Trkman, and S. Afshari. 2019. The Role of Business Analytics Capabilities in Bolstering Firms' Agility and Performance. *International Journal of Information Management* 47: 1–15.

Carroll, N., M. Helfert, and T. Lynn. 2013. *A Contingency Model for Assessing Cloud Composite Capabilities*. CLOSER, 515–519.

———. 2014. Towards the Development of a Cloud Service Capability Assessment Framework. In *Continued Rise of the Cloud*, 289–336. London: Springer.

Cisco. 2013a. Internet of Everything: A \$4.6 Trillion Public-Sector Opportunity. https://www.cisco.com/c/dam/en_us/about/business-insights/docs/ioe-public-sector-vas-white-paper.pdf.

———. 2013b. Embracing the Internet of Everything to Capture Your Share of \$14.4 Trillion. https://www.cisco.com/c/dam/en_us/about/business-insights/docs/ioe-economy-insights.pdf

Cordoso, J. 2019. *The Application of Deep Learning to Intelligent Cloud Operation*. Paper presented at Huawei Planet-scale Intelligent Cloud Operations Summit, Dublin, Ireland, 1 November 2019.

Côrte-Real, N., T. Oliveira, and P. Ruivo. 2017. Assessing Business Value of Big Data Analytics in European Firms. *Journal of Business Research* 70: 379–390.

Dang, Y., Q. Lin, and P. Huang. 2019. *AIOps: Real-World Challenges and Research Innovations*. Proceedings of the 41st International Conference on Software Engineering: Companion Proceedings, 4–5. IEEE Press.

Eijkhout, V., R. van de Geijn, and E. Chow. 2016. *Introduction to High Performance Scientific Computing*. Zenodo. https://doi.org/10.5281/zenodo.49897

Eivy, A. 2017. Be Wary of the Economics of "Serverless" Cloud Computing. *IEEE Cloud Computing* 4 (2): 6–12.

Ezell, S.J., and R.D. Atkinson. 2016. *The Vital Importance of High Performance Computing to US Competitiveness*. Washington, DC: Information Technology and Innovation Foundation. Accessed 23 October 2017. http://www2.itif. org/2016-high-performance-computing.pdf

Ferreira, L., E. da Silva Rochay, K. Monteiroy, G. Santos, F. Silva, J. Kelner, D. Sadok, C. Bastos Filho, P. Rosati, T. Lynn, and P. Endo. 2019. *Optimizing Resource Availability in Composable Data Center Infrastructures.* Proceedings of the 9th Latin-American Symposium on Dependable Computing. IEEE.

Gillen, A., C. Arend, M. Ballou, L. Carvalho, A. Dayaratna, S. Elliot, M. Fleming, M. Iriya, P. Marston, J. Mercer, G. Mironescu, J. Thomson, and C. Zhang. 2018. *IDC FutureScape: Worldwide Developer and DevOps 2019 Predictions.* IDC.

Glikson, A., S. Nastic, and S. Dustdar. 2017. *Deviceless Edge Computing: Extending Serverless Computing to the Edge of the Network.* Proceedings of the 10th ACM International Systems and Storage Conference, 28. ACM.

Haller, S., S. Karnouskos, and C. Schroth. 2009. *The Internet of Things in an Enterprise Context.* Future Internet Symposium, 14–28. Berlin, Heidelberg: Springer.

Hendrickson, S., Sturdevant, S., Harter, T., Venkataramani, V., Arpaci-Dusseau, A. C. and Arpaci-Dusseau, R. H. (2016). Serverless Computation with OpenLambda. *Elastic* 60: 80.

Iorga, M., L. Feldman, R. Barton, M. J. Martin, N. S. Goren, and C. Mahmoudi. 2018. Fog Computing Conceptual Model (No. Special Publication (NIST SP)-500-325).

Lehrer, C., A. Wieneke, J. vom Brocke, R. Jung, and S. Seidel. 2018. How Big Data Analytics Enables Service Innovation: Materiality, Affordance, and the Individualization of Service. *Journal of Management Information Systems* 35 (2): 424–460.

Liu, F., J. Tong, J. Mao, R. Bohn, J. Messina, L. Badger, and D. Leaf. 2011. NIST Cloud Computing Reference Architecture: Recommendations of the National Institute of Standards and Technology (Special Publication 500–292).

Lynn, T. 2018. Addressing the Complexity of HPC in the Cloud: Emergence, Self-organisation, Self-management, and the Separation of Concerns. In *Heterogeneity, High Performance Computing, Self-Organization and the Cloud,* 1–30. Cham: Palgrave Macmillan.

Lynn, T., P. Rosati, A. Lejeune, and V. Emeakaroha. 2017. *A Preliminary Review of Enterprise Serverless Cloud Computing (Function-as-a-Service) Platforms.* 2017 IEEE International Conference on Cloud Computing Technology and Science (CloudCom), 162–169. IEEE.

Lynn, T., P. Rosati, and P. Endo. 2018. *Towards the Intelligent Internet of Everything: Observations on Multi-disciplinary Challenges in Intelligent Systems.* Proceedings of the Research Coloquio Doctorados: Tecnología, Ciencia y Cultura: una visión global.

Marinescu, D.C. 2017. *Complex Systems and Clouds: A Self-organization and Self-management Perspective.* Cambridge, MA: Elsevier.

Mell, P., and T. Grance. 2011. *The NIST Definition of Cloud Computing*, Special Publication 800–145. Gaithersburg, MD: National Institute of Standards and Technology.

Mikalef, P., J. Krogstie, I.O. Pappas, and P. Pavlou. 2019. Exploring the Relationship between Big Data Analytics Capability and Competitive Performance: The Mediating Roles of Dynamic and Operational Capabilities. *Information & Management* 57: 103169.

Nadkarni, A. 2017. *Quantifying Datacenter Inefficiency: Making the Case for Composable Infrastructure*. IDC.

Prasad, P., and C. Rich. 2018. *Market Guide for AIOps Platforms*. Gartner.

Rosati, P., G. Fox, D. Kenny, and T. Lynn. 2017. *Quantifying the Financial Value of Cloud Investments: A Systematic Literature Review*. 2017 IEEE International Conference on Cloud Computing Technology and Science (CloudCom), 194–201. IEEE.

Schryen, G. 2013. Revisiting IS Business Value Research: What We Already Know, What We Still Need to Know, and How We can Get There. *European Journal of Information Systems* 22 (2): 139–169.

Scogland, T.R., C.P. Steffen, T. Wilde, F. Parent, S. Coghlan, N. Bates, et al. 2014. *A Power-Measurement Methodology for Large-Scale, High-Performance Computing*. Proceedings of the 5th ACM/SPEC International Conference on Performance Engineering, 149–159. ACM.

Shan, A. 2006. Heterogeneous Processing: A Strategy for Augmenting Moore's Law. *Linux Journal* 2006 (142): 7.

Yeo, S., and H.H. Lee. 2011. Using Mathematical Modeling in Provisioning a Heterogeneous Cloud Computing Environment. *Computer* 44 (8): 55–62.

INDEX

© The Author(s) 2020
T. Lynn et al. (eds.), *Measuring the Business Value of Cloud Computing*, Palgrave Studies in Digital Business & Enabling Technologies, https://doi.org/10.1007/978-3-030-43198-3